Kamil Satarovich Ashimov
Galina Mikhailovna Chernova

Biomorphological peculiarity of adaptation of the true pistachio

Kamil Satarovich Ashimov
Galina Mikhailovna Chernova

Biomorphological peculiarity of adaptation of the true pistachio

Monograph

ScienciaScripts

Cover image: www.ingimage.com

This book is a translation from the original published under ISBN 978-3-659-86566-4.

Publisher:
Sciencia Scripts
is a trademark of
Dodo Books Indian Ocean Ltd. and OmniScriptum S.R.L publishing group

120 High Road, East Finchley, London, N2 9ED, United Kingdom
Str. Armeneasca 28/1, office 1, Chisinau MD-2012, Republic of Moldova, Europe
Managing Directors: Ieva Konstantinova, Victoria Ursu
info@omniscriptum.com

Printed at: see last page
ISBN: 978-620-8-54548-2

CONTENTS.

INTRODUCTION

Intensification of industrial (garden) culture of pistachio in Uzbekistan predetermines the use of new technological methods that allow the most effective involvement of agricultural turnover of low-productive land in the valley parts of the foothills, especially on poor sandy loamy pebble soils of the cones of removal of numerous mountain rivers. They are usually represented by non-soil formations - uncolmatized boulder-pebbles and pebbles of different degrees of calcification with colmatized fine-grained horizon up to 0.5m and make up large areas of semi-desert stony spaces with sparse wormwood-solanaceous vegetation. Only in Fergana, Namangan and Andijan regions they occupy about 400 thousand hectares in areas with intense wind activity (e.g. Kokand group of districts, including the site "Besh-Aryk" in the valley part of the foothills of the Turkestan ridge), where the research was conducted.

This drought-resistant and relatively frost-resistant species with a well-developed root system and high adaptation to different natural and climatic conditions, as evidenced by the growth of pistachio in rainfed foothills of almost all ranges of Asia, including Uzbekistan. It is interesting to note that only in the Fergana Valley the slopes of Chatkal, Turkestan and Alai ridges were covered with pistachio thickets in the past, where only small pockets of pistachio trees have been preserved to date, and mostly in places difficult for people to access. It is not without reason that the first explorers of the nature of Turkestan region in the 18th century called the Fergana Valley "the country of pistachio". It is important to involve into agricultural turnover not only the rainfed foothills previously occupied by pistachio, but also practically undeveloped "abandoned" pebble lands. Long-term experience of forest scientists of Central Asia and especially fundamental researches of T.A.Zheltikova (1977) on production of forest crops on zakolmatized pebbles of cones of outcrop showed that development of pebble lands that were earlier considered as "abandoned" on the basis of irrigation and melioration can solve the problem of increasing their soil fertility and especially will open perspectives for their development for agricultural and fruit crops. It has been revealed that changes occurring in soil under the forest or garden canopy are expressed in the increase of humus horizon thickness and humus content, improvement of soil structure and water-physical properties. This, in turn, opens prospects for the development of pebble lands for pistachio plantations. Due to high adaptation to various natural and climatic conditions inherent to this breed, the true pistachio is one of the most promising breeds for involvement in agricultural turnover of hard-to-recover pebble lands. Therefore, restoration of the former pistachio area in the Fergana Valley and return of pistachio to its homeland is an important state task. Taking into account the growing water deficit in this region and availability of huge reserve of vacant lands suitable for cultivation of exceptionally drought-resistant pistachio, which is not demanding to conditions, will allow to restore not only the former areal of this breed, but also to get valuable pistachio products and what is important to improve the ecological situation in this region as a whole. In connection with the set state task, cultivation of industrial plantations of pistachio on a varietal basis, carrying out zoning of valuable assortment, will help to raise stable employment of the population, improve their welfare. The same problem is valid practically for all regions of the Republic of Uzbekistan, as well as other regions of Central Asia. Justification (on the basis of) biologo-ecological adaptive capacity of pistachio allows hoping for intensification of involvement in agricultural turnover of many empty rainfed foothills and low mountains by growing plantation (garden) crops of pistachio.

Figure 1: Pistachios of Andijan, Republic of Uzbekistan

CHAPTER 1. The importance of the true pistachio

The importance of the real pistachio is great. On the one hand, it is the main forest-forming species in arid foothills and low mountains in almost all ranges of the Central Asian region, playing, due to its well-developed root system, a huge soil-protective and water-protective role. Pistachios grow at altitudes from 500 to 2000m above sea level, being an excellent species for maintaining a healthy ecosystem. Pistachio fruits particularly well at altitudes between 800 and 1300m above sea level. Due to its high flavor qualities, pistachio fruits are valued 3-4 times more expensive than walnut and almond fruits on the world market. Both directly in raw form and after various processing, they are used in the confectionery industry, in the production of higher grades of sausages, as well as dietary products.

In the Canon of Medical Science, compiled by Abu-Ali Ibn Sina, it is given a significant place in the treatment of liver diseases, stomach, as a means of healing external wounds. Pistachio is used in the treatment of chronic lung diseases. Tincture of pistachio pericarp is drunk for gastric diseases. However, the pistachio tree can also be a source of tannins, medicinal tannin and resin. For thousands of years, pistachios in the East have been considered an excellent remedy for removing toxins from the body.

Pistachio resin, called pistachio turpentine, is highly technical and is suitable for the manufacture of alcohol and oil varnishes, widely used in aircraft construction.

It is not without reason that the pistachio is called a combine tree, as all its parts (wood, fruit, resin) can be used by people for their needs. In the countries of the Central Mediterranean basin (Iran, Turkey, Syria, etc.) pistachio is called "green gold" or "golden tree" because of the high income it brings. Moreover, in these countries, the main production is obtained not from wild plants, but from garden plantations.

It is necessary to emphasize the exceptional importance of pistachios for people living in arid arid foothills, where practically no other species can grow without additional irrigation. The pistachio is not afraid of dry hot winds ("Garmsili"); it is resistant to heat and drought. All this characterizes it as an exceptionally resistant, adapted species, which not only improves the microclimate of the adjacent territories, but also has a great soil-improving and water-protecting value in the pistachio zone.

But first of all, the real pistachio is a valuable "nutcracker", the fruits of which are valuable foodstuffs and irreplaceable raw material for food, vitamin and a number of other industries. Pistachio seeds, being a high-calorie dietary product, contain from 40 to 60% and more fats, 15-20% proteins, 3-8% sugars and many microelements. In them from carbohydrates are found glucose, fructose, sucrose, raffinose (V.F. Shcheglova et al., 1975), and from nitrogen-containing compounds - leucine, phenylalanine, valine, threonine, arginine, lysine, glutamine and asparagine (E.Y. Babekova, 1979). Due to high flavor qualities of seeds, fruits of pistachio are valued 3-4 times more expensive than walnut and almond fruits in the world market (V.P. Alekseev, 1963). Pistachio nuts are used both in fresh form and in the food industry for making pistachio oil, many confectionery and culinary products.

The pistachio has many other valuable qualities. The gall formations on the leaves, known in the literature as "buzgunch", contain up to 50% tannins, which are highly valuable raw materials for the pharmaceutical industry. According to K.P. Popov (1979), the leaves themselves contain up to 18% of tannins, which are used to produce dyes for coloring fabrics in different shades.

The true pistachio (the only representative of the numerous genus Pistacia) is a valuable nut-bearing tree with edible fruits, so called pistachio nuts, which have received worldwide recognition due to a complex of valuable properties. Since ancient times, pistachio nuts have been considered a delicacy by the population of the Eastern countries, and the tree bearing these fruits is a "golden tree" or "green gold", as they are not only a source of their income, but also their livelihood.

The habitat of wild pistachio in Central Asia (Uzbekistan, Tajikistan, Turkmenistan, Southern Kyrgyzstan, Southern Kazakhstan) is confined to the most warmed mountain slopes and foothills of practically all ridges with accumulative form of relief with the cover of gray soils developed on loess.

In this connection, we consider it necessary to emphasize that characterized by exceptional drought tolerance, heat resistance and relatively high frost resistance, the true pistachio in the process of evolution has adapted to the diverse natural and climatic conditions of the Central Asian region.

All this characterizes the pistachio as an exceptionally plastic and adapted species and opens great prospects for its introduction into culture in different natural-ecological conditions of Central Asian republics. It is also important that the pistachio is not only a valuable nut-bearing plant, but also plays a huge soil-protective and water-protective role in the zone of semi-desert foothills in the south of Central Asia, improving the microclimate of adjacent areas.

Analyzing the above stated, it is necessary to emphasize that pistachio represents the most valuable natural phenomenon in the arid zone of the south of Central Asia. Preservation and replenishment of this species in its original homeland - Central Asia, in the complex of solving the problems of conservation and strengthening of nature protection functions of the real pistachio in mountainous and foothill areas, is of great importance for the whole Central Asian region.

CHAPTER 2. The distribution area of the true pistachio in Central Asia.

(re)The genus Pistacia vera L- pistachio in the family Anacardiaceae (Anacardeacea Linne) comprises about 20 species of small evergreen and deciduous trees and shrubs distributed mainly in subtropical and tropical regions of the northern hemisphere.

Only one species grows in the Central Asian republics - the true pistachio (or edible or noble) - Pistacia vera L. The true pistachio was described by C. Linnaeus in 1733, hence Pistacia vera Linne.

The natural distribution area of the true pistachio in Central Asia is confined to the low mountains of the Tien Shan, Pamir-Alai and Kopetdag mountain systems and is characterized by a rather considerable extent in the system of geographical coordinates - from the foothills of the Kyrgyz Range ($43^{0}13^{*}$ N$^{)\ to\ the}$ foothills of Paropamiz $^{(35(0)5(*)\ S)}$from north to south from the Boam Gorge of the Kyrgyz Range foothills ($35^{(0)5(*)\ S)}$.N) to the foothills of Paropamiz ($35^{0}5^{*}$ S) from north to south, from the Boam Gorge of the Kyrgyz Range foothills ($75^{(0)}$45 E) to South-West Kopetdag ($55^{(0)}4^{*}$ W) from east to west (Bulychev 1969; Kamelin, 1973). According to K.P.Popov (1979), the extent of pistachio distribution from north to south is up to 800 km, from east to west - about 1300 km. The altitudinal range of pistachio distribution is also wide, from 500 (600) to 1800 (2000m) above sea level.

There is reliable information from archaeologists (Lisitsina, 1972) that in the Stone Age (more than 1000 years ago) the area of pistachio in Central Asia occupied more than 2 million ha, although in modern times, unfortunately, due to sometimes unreasonable human activities (cutting, grazing, etc.) it does not exceed 250 thousand ha. Only for the past few decades in Uzbekistan the area of natural pistachio thickets has decreased from 70 thousand ha to 12 thousand ha.

The confinement of pistachio to certain mountain systems separated from each other and some differences in natural and climatic conditions, inherent in them, allows us to distinguish three main regions of natural distribution of pistachio in Central Asia:

1. Tien Shan (northern) - includes pistachios (about 30 thousand ha) of Southern Kyrgyzstan, growing in the foothills of Fergana and Changyr-Tash ranges within the relative altitudes from 800 to 1600m above sea level, which are confined to desert-low and adyr geomorphologic landscapes. The same region, in the form of small "islands" and sparse stands, includes pistachios of northern Kyrgyzstan (foothills of the Kyrgyz Range), Uzbekistan (Pskem and Chatkal Ranges), southern Kazakhstan (Talas Alatau and Karatau), northern Tajikistan (Kuramin Range).

2. Pamir-Alai (central), which includes pistachios of southern Tajikistan (about 115 thousand ha) southern Uzbekistan (about 12 thousand ha) and eastern Turkmenistan (5 thousand ha). In Tajikistan, pistachio grows along the following ridges: Pripyanji Karatau, Tereklitau, Gazimailik, Aruktau, Sarsaryak, Chaltau; in Uzbekistan - Babatag ridge; in Turkmenistan - Kushtagtau ridge, with a lower altitudinal limit of 500 (600) m above sea level and an upper limit of 1500 (1600) m above sea level.

Figure 2: Tereklitau Ridge, Tajikistan

3. Kopetdagh (southern), includes pistachio southwest Turkmenistan, under the general name of Batkhyz (Kushkinskaya Grove, Pul-Khatum Grove, Badkhyz Reserve). Pistachio trees here (about 75,000 ha) are rather sparse (no more than 40 trees per ha), growing on the so-called "hills" of the foothills of Paropamiz and Kopetdag, with altitudinal elevations of 600-1200m above sea level.

The climate of all three regions has a clearly pronounced continental character, with sharp seasonal and daily fluctuations of meteorological elements. Absolute maximum air temperatures (July) can reach +48^{0}C, absolute minimum in the coldest month (January) drops to -30^{0}C, and even to -40(^{0})C along the northern border of the area (Table 1).

Table 1.

Some elements of climate of the pistachio growing area in Central Asia (700-1300m.a.s.l.)

№	Region	Annual precipitation, mm	Absolute t° air		Average bottom t^0 air	Monthly averages t^0 air	
			poppy	min		January	July
1	Tien Shan:						
	Northern Kyrgyzstan	418	38	-40,1	10,3	-3,6	24,4
	South Kyrgyzstan,	400	39	-28,0	13,3	-0,6	26,6
	South Kazakhstan	686	41	-30,0	11,7	-2,9	25,6
2	Pamir-Alai:						
	South Tajikistan	436	44	-27,4	14,7	-0,3	28,7
	Southern Uzbekistan	360	47	-25,0	16,0	+3,0	29,2
3	Kopetdagh Badkhyz	289	48	-33,0	14,5	+3,5	28,1

Mean monthly perennial air temperatures in seasonal dynamics range from +24.4-26.6^{0}C (July) to -0.3 +3.5$^{(0)}$C (January), with a mean annual air temperature for the three regions+ of 13.2$^{(0)}$C (from 10.3^{0}C at the northern boundary of the range to 16.00C - at the southern boundary).

And as the data of Table 2 on temperature gradient and annual precipitation sum testify, the northern, central and southern limits of distribution of this rock differ from each other.

Table 2

Climatic conditions in the pistachio belt in Central Asia in the Tien Shan, Pamir-Alai and Badkhyz.

Observation points of meteorological stations	Average annual precipitation, mm	Average annual air temperature 0C	Absolute air temperatures		Authors
			max.	minimum.	
1.Kara-Archa (Kyrgyz ridge near Almaly tract).	390	7,7	40,1	-40,1	A.S.Bulychev (1970).
2. Jalal - Abad (Fergana Range)	350-403	13,1	41,0	-28,0	V.E.Ozolin (1969).
3. Northern part of the Babatag Ridge	360	16,0	47,0	-25,0	E.I.Moskvina (1968)
4.Gandjina (Southern Tajikistan Aktau Ridge)	339	14,5	43,0	-20,5	Tables of weather stations for 1953- 1972.
5. Kushka (southern Turkmenistan, Badkhyz)	289	14,4	48,0	-33,0	Reference book on climate of the USSR, issue 30 1966-1969

It is evident from the data of the tables that the climate in the north and south within the range is sharply continental. The true pistachio, being a subtropical plant, withstands frosts up to -40^0 in the northern Tien-Shan. According to A.S. Bulychev (1970), no traces of subfreezing are observed in it. At the same time, this tree withstands summer temperatures up to $47^0 48^0$C. The sum of precipitation gradually decreases from north to south and does not reach less than 300mm per year in the south. From June through September, the soil surface warms to 7075^0. In summer, as the daily temperature increases, the relative humidity decreases to 12-15%, which leads to a sharp increase in evapotranspiration.

The main environmental factor limiting the condition of pistachio in the low mountains and foothills of southern Central Asia is temperature regime and soil moisture. High summer temperatures do not cause noticeable suppression of growth and development of heat-resistant and heat-loving pistachio. In general, the reduction of vitality of this breed is observed in the summer period when the south-eastern winds from Afghanistan, the so-called (Afghans), which have a drying effect on the climate of the southern parts of the region and in other years cause burns of unformed fruits. Late spring frosts also cause certain damage to pistachio, which are especially destructive during the period of its mass flowering (1-2 decade of April).

It has been established that for normal growth and development of pistachio throughout the whole region of its growing in Central Asia the sums of stable average daily air temperatures above +5^0C not less than 3400^0; above +$10^{(0)}$C not less than 3200; above +20^0C not less than 2000-$2200^{(0)}$C are necessary. The average duration of the vegetation period should be 200 or more days, and frost-free period - not less than 160 days.

It is the above temperature gradients that restrain the growth of pistachio at high hypsometric elevations (above 1400m.a.s.l.), especially the formation of yield and cause its sporadic growth at the northern border of its range (southern Kazakhstan and northern Kyrgyzstan)- (Table 3).

Table 3 Thermal indices of the growing season of pistachio in its natural growing area in Central Asia.

Wysocki- ta m	Sum of positive average daily temperature p	Duration of the period with temperatures above

n.u.m.	+50	+100	+200	+50		+100		+200	
				Cf. dates of transition a	Days elapsed	Cf. dates of transition a	Duration in days	Cf. transition dates	About a day x
500 700	5810	5565	4110	18,02 30,11	287	24,0209,11	258	08,05 - 15,10	160
800 1000	5020	4760	3130	22,02 12,11	263	15,03 30,11	229	17,05 - 28,09	134
1100 1300	4090	3890	2130	16,03 10,11	239	12,04 27,10	198	08,06 - 10,09	94
1400-	3650	3510	1940	24,03-	221	13,04-	189	08,06	93
1600				31,10		9,10		- 09,09	
More than 1,600	3020	2520	1230	05,04 30,10	209	15,04 09,10	177	15,06 - 04,09	81

Annual precipitation, reaching an average of about 350mm, gradually decreases from north-east to south-west, from 418mm in the foothills of the Kyrgyz Range to 290mm in Badkhyz. Changing within geographical coordinates, the annual precipitation at the same time, within each region, varies depending on the vertical profile, gradually increasing as the terrain rises above sea level - from 200-250mm at altitudes of 500600m (Nurata Range, Uzbekistan), to semi-desert foothills of Paropamiz in Turkmenistan and up to 450 (600mm) at altitudes of 1300 and more meters above sea level (Rangentau Ridge in Tajikistan, Kirghiz Ridge in northern Kyrgyzstan).

Seasonal dynamics of precipitation is also very peculiar. Their maximum (up to 88%) falls in the winter-spring period. March and April are the rainiest, accounting for about 35% of the annual precipitation. In summer, the hottest period, they are practically absent.

Soils of the pistachio area of Central Asia, with the exception of chestnut and brown-carbonate soils of the Kyrgyz ridge, are mainly of the serozem type. Water regime of gray soils in the pistachio zone corresponds to the zone of extremely insufficient moistening. Hence, it is clear what great importance in plant biocenosis in these conditions belongs to drought-resistant pistachio, which, having great adaptive flexibility, adapts well to various types of soils, different in mechanical composition, structure, physical properties, aeration. However, being well adapted to harsh soil conditions, pistachio does not tolerate excessive salinization and overwatering. It prefers light sandy loam and loamy, well-drained soil types. It grows very well on typical and dark sierozem soils. Resistance of pistachio is expressed not only in the fact that it tolerates the effects of elevated (up to $+40^0$C) air temperatures, but also that it does not suffer from low (up to - 40^0C) air temperatures during the winter dormancy period on the northern border of its range.

Pistachio trees of Central Asia are characterized by sparse canopy and closed root systems. The drier the conditions, the less moisture in the soil, the further the root systems extend in the upper soil horizon, providing the tree with normal growth and development the dry summer period. Therefore, it is not accidental that in the extreme south of the area in Badkhyz, in the most extreme conditions of pistachio growth in terms of moisture availability, the stand density is 0.1-0.2 (30-40 trees per hectare), while in the Pamir-Alai and Tien Shan regions, with more moderate hydrothermal regime, the density of pistachio stands increases to 0.3-0.4 (70-80 trees per hectare). In small isolated areas with higher moisture content, up to 120-150 trees per hectare grow.

Analyzing the above, it should be noted that the true pistachio is a valuable natural phenomenon in the arid zone of southern Central Asia (re) Pistachio is also cultivated on the Apsheron Peninsula in Crimea (Voinov, 1948) in Moldova, Transcaucasia, Georgia, Kakheti, Dagestan (Fedorov, 1957). The oldest pistachio trees are indicated for the Crimea, where it was introduced in the late 18th and early 19th centuries. Of the European countries cultivating pistachio, Italy ranks first, where it is cultivated on the islands of Sicily and Sardinia. Pistachio is also cultivated in France, Spain, Portugal and Greece.

In North Africa, cultivated pistachio is cultivated mainly in Tunisia and Algeria. Pistachio grows widely in Syria, Libya, Messopotamia, Iraq, California, Mexico, Texas, Arizona (Belyaeva, 1938; Lemaistre, 1959).

Conclusions: **-Ecological** conditions of growing pistachio true over all three regions of Central Asia are characterized by sharply continental climate, where maximum air temperatures reach +48°C, and minimum to -30°C, and along the northern border to -$40^{(o)}$C. Average monthly temperatures in July are +24.4 -26.6°C, with an average annual air temperature across the three regions of +13.2°C, and the soil surface is very warm to 70-75°C

- Relative humidity decreases under these environmental conditions to 12-15%, resulting in a dramatic increase in evapotranspiration.

- For the southern regions, the decrease of pistachio viability is real during the hot summer period, when winds from Afghanistan blow so-called "Afghans", which have a drying effect on the climate.

- The main ecological factor limiting the condition of pistachio in the low mountains and foothills of southern Central Asia is the temperature regime and soil moisture.

- Late spring frosts are very dangerous during the period of mass flowering of pistachio (I and III decades of April).

- In all three regions of Central Asia, for normal growth and development of pistachio, the sums of stable average daily air temperatures above +5°C not less than -$3400^{(o)}$C; above +$10^{(o)}$C not less than 3200°C; above +20°C not less than 2000-$2200^{(o)}$C are necessary.

- An important ecological feature characteristic of the whole of Central Asia is the annual precipitation of 350mm on average, which gradually decreases from north-east to south-west - from 418mm in the foothills of the Kyrgyz Range to 289mm in

Badkhyz.

- The annual amount of precipitation within each area varies depending on the vertical profile. At lower altitudinal ranges (500600m) the average annual precipitation is 200-250mm, - (foothills of the Nurata Range), 450-600mm at altitudes of 1300 and more meters above sea level (Rengentau Range in Tajikistan, Kirghiz Range in Northern Kyrgyzstan).

- The ecological characteristics include different soil conditions of pistachio in the three regions. Thus, the soils of the Kyrgyz ridge are chestnut and brown carbonate, while in other regions the soils are mainly of sierozem type. Pistachio has adaptive capacity and adapts well to a variety of soil types, different in mechanical composition, structure, physical properties and aeration. Well adapted to harsh soil conditions, pistachio does not tolerate excessive salinity and overwatering at all, it prefers light loamy and loamy, well-drained soils of sierozem type(repeat).

Thus, based on the ecological peculiarities of pistachio growing in Central Asia, it can be stated that in the north of the range boundary in Southern Kyrgyzstan it grows only from 800m to 1600m altitude; in the Pamir-Alai (Central) region - Southern Uzbekistan, Southern Tajikistan - pistachio grows from 500m (600m) to 1800m altitude, and in the Kopetdag (Southern region), which includes pistachios of south-western Turkmenistan, under the common name Badkhyz, it grows at altitudes from 600 to 1200m.

The wide geographical and altitudinal range of the true pistachio within Central Asia, first of all, testifies to the high adaptation of this persistent species to grow in different natural and climatic conditions of the Central Asian region and, consequently, to the unlimited territorial possibilities of its introduction into culture.

CHAPTER 3. Morphology and biology of the true pistachio tree

Pistachio is a low tree from 3 to 4 (rarely 5-7m) tall, mostly multi-stemmed, consisting of numerous trunks of different age and diameter. Its crown has the shape of a large bush, like all light-loving trees, is rather loose, the branching type is sympodial. Each tree usually has three trunks, with average trunk diameter in middle-aged stands (60-80 years) from 17 to 25 cm. At relatively low height of the tree the crown reaches 4-5m in diameter. Trunk bark of old branches is longitudinally - cracked, brownish-gray; of young branches - broad, ashy-gray; on annual branches and shoots - smooth, reddish-brown. Vegetative buds 0.6-0.7 cm. long, broadly ovate, brown, with reddish tinge. Scales broadly ovate, up to 0.5 cm. long, strongly pubescent in the middle and along the edges. The morphological indicator as the structure of pistachio leaves is very important. Leaves are complex in shape - three to five leaflets, with one, two (three) leaflets as exceptions. In early spring at the beginning of vegetation the leaves are thin, very delicate, later they become thick, leathery. Leaflets are mostly round-ovate, less often oblong, rounded or broadly lanceolate, sometimes almost square, permanently acuminate at the apex, with a short sharp end, less often rounded and truncate, seven to eight cm long, five to six cm wide, sometimes the leaflet width is almost equal to the length or, on the contrary, twice less than it. The base is rounded or broadly wedge-shaped, sometimes gradually turning into a petiole. Leaflets are almost sessile from above, especially along the veins, shortly hairy. In the period of leaf unfolding they are densely hairy, woolly along the edge of the veins. The anatomomorphological structure of pistachio leaf reflects the growing conditions of this species and characterizes its adaptive properties to environmental factors. In extreme conditions of moisture availability, semi-desert foothills of the area, where annual precipitation does not exceed 200mm, the pistachio leaf is simple and consists of one (rarely three) small leaflets 3-4cm long and 2-3cm wide. In the best conditions, the pistachio leaf is complex, consisting of five, even seven large leaflets up to 15cm long and 12cm wide. The leaf mesophyll consists of five to seven rows of palisade parenchyma, spongy parenchyma is absent at all, which indicates high heliophilicity of the pistachio (it is a plant of sunny habitats), so the pistachio has a sparse crown, illuminated on all leaves.

The petiole is densely hairy at the beginning of vegetation, even woolly, rather thick.

Chlorophyll is distributed evenly throughout the mesophyll cells. The upper and lower epidermis consist of tall isodiametric thick-walled cells with a thick cuticle, which makes leaves rigid. Stomata and hairs are evenly distributed on both sides of the epidermis. The hairs are unicellular, glandular. According to E.I. Adamovich and B.N. Norin (1954), the leaf veins have resinous passages that pass into the stigma.

The beginning of pistachio bud swelling falls on the first decade of March. In the first days of April they open. At the end of the first decade of April leaves begin to appear and only by the end of April or early May they unfold completely. Fall foliage begins in November.

Pistachio is a dicotyledonous plant, with unisexual flowers arranged in compound panicles. The pistillate flowers are arranged in loose oblong-ovate panicles, up to six to eight cm long and two to four wide. The panicles with staminate flowers are compact, oval-rounded, with a length of four to six centimeters and a width of three to five centimeters. Staminate flowers with a simple perianth of two to six oblong, very thin leaflets covered with white short downy hairs over the entire surface. Stamens are three to five with large anthers. The pollen grains of pistachio are flattened-spheroidal, with a short polar axis, 32.4 - 36 mm with rounded or broadly oval pores in the number of 5-6 (7). Pore edges are smooth, slightly thickened. The grains on the surface of the pore membranes are collected in a rounded or oblong tectum (Kupriyanov,1961). Pistillate flower consists of five lanceolate to oblong, thin, almost filmy perianth leaflets covered with short downy hairs over the entire surface. The pistil is 3mm long, the ovary is rounded, and the stigma is three-divided, with lamellate, downwardly bent leaves.

Flower buds are laid on the shoots of the current year in late May or early June after the cessation of growth in the year preceding flowering. The buds become fully formed by winter. Flowering of pistachio begins in April. Such late flowering is caused by the pistachio's great sensitivity to low temperature, especially during budding, although in the state of winter dormancy it can tolerate frosts of -30°C without any damage

(Blinovsky et al., 1957). Flowering of male and female trees is observed with an interval of 5-6-10 days. The timing of flowering onset can vary up to 5-8 days depending on climatic conditions of the year. The flowering period of one tree with female flowers is from 5 to 9 days, with male flowers it is usually shorter: 5-7 days. A greater number of male trees in the plantation with a greater amplitude of flowering time 17 ensures normal pollination. The flowering period of one flower, counting from the beginning of stigma lobes unfolding to their fading - from 24 to 60 hours.

Fruits are collected in bunch-like clusters 12-22 cm long and 6-12 cm wide. The fruit bones are on a thick and rather long peduncle and consist of pericarp, nutritive protein and embryo. The pericarp is subdivided into epicarp, mesocarp and endocarp. According to V.A. Nassonov (1934), the epicarp (outer sheath) has all similar features to the leaf epidermis, differing from it only by the fact that the cell walls are thinner, the stomata are larger and there are fewer of them per unit area. Mesocarpy (fruit sheath pulp) is a vascular-fiber bundle with strongly reduced vascular system and developed excretory passages in phloem. At the beginning of ripening, the mesocarp is juicy and resinous; at full ripening it dries out. The endocarp (hard bony sheath) or pith consists of stony cells.

The seed consists of two green, very fleshy achenes with the embryo between them. At full maturity, the epicarp and mesocarp fall off and the endocarp opens at the suture. The pistachio is often divided into "opening" and "non-opening" pistachios based on this characteristic. The degree to which the endocarp opens is usually considered a varietal trait, but as Spinae and Penoisi (Spinae,Penoisi 1957) show, this trait can be regulated by appropriate tillage, fertilization, and placement in appropriate natural areas. Pistachio nuts are very variable in shape and size. Most often they are roundly elongated and even curved with a rounded or cone-shaped top. In color pericarp in ripe and ripe fruit can have up to 15 shades (from white, pink to bluish. The length of the stone, with fluctuations of 1.2 - 2.5 cm, on average is equal to 1.9 - 2.0 cm, width with fluctuations of 0.9 - 1.4 cm, on average varies from 0.4 to 1.5 g, the most common nuts 0.7 - 0.8 g. According to E.E. Kern (1931), the pistachio fruits from which the males are supposed to grow are placed on top of the inflorescences. The yield depends largely on the age and conditions in which the pistachio grows. According to foreign authors (Avanzato D, Monastra F. 1982, 1983, etc.), trees aged 100 years and more, even grown in orchards, yield no more than 20 kg. In natural pistachio forests of Babatag (Uzbekistan), according to I.K.Trosko (1937, 1940), the yield of middle-aged trees (50-70 years old) usually does not exceed 1-3 kg per tree. The most productive years are the years from the second to the third. In natural conditions pistachio grows very slowly. It reaches a height of 1m in 3-5-7 years, and 2m in 5-25 years. When studying pistachio root systems, the vertical root grows intensively in the first year. By the time the leaves appear on the soil surface (end of March), the root manages to penetrate to a depth of 20-25cm. A month later, the roots penetrate to a depth of 40-50cm. By the end of the first year, when the height of the above-ground part of the seedling is no more than 9 - 10 cm the root deepens to 100 (150) cm. In the first 3 - 4 years with very slow growth of pistachio in height roots continue to deepen and branch; by 5-6 years their length is equal to 3m, from 15-20 years horizontal growth of roots is well marked.

Thus, the anatomomomorphological structure of the pistachio leaf reflects the growing conditions and characterizes its adaptive properties to climatic conditions. Under the harshest conditions, in the semi-desert foothills of the range, where annual precipitation does not exceed 200 mm, the pistachio leaf is simple and consists of one (rarely three) small leaflets 3-4 cm long and 2-3 cm wide. In the best conditions with annual precipitation of 300-450 mm, the pistachio leaf is complex and consists of 5 or even 7 large leaflets up to 15 cm long and 12 cm wide.

Stomata of pistachio are located on the upper and lower sides of the leaf and their adaptive properties are their wide opening during the whole daylight period, as well as their huge number, from 250 to 450 pcs/cm^2, also depends on different places of growth, adapting to the given conditions. Another adaptive property of pistachio by morphological traits is its flowering with different intervals - in 5 -6 - 10 days. Biological and morphological peculiarity of pistachio is the development of a powerful root system, which already in the first year grows and reaches a depth of 100 (150) cm, and then horizontal roots with a huge network of small sucking roots develop. This adaptability of pistachio to develop a powerful root system in different soil and climatic conditions is one of the main adaptive properties, as it creates a huge area of moisture supply for this unique species throughout Central Asia.

CHAPTER 4. Evidence on plant resistance and adaptation drought (literature review)

A major stage in the study of plants in arid habitats is associated with the works of N.A.Maksimov "On the issues of drought resistance of plants, - writes N.A.Maksimov, (1927, 1929-1937) we do not have any summaries at all". The author collected and systematized information on this problem, which is presented in the monograph "Physiological bases of drought resistance of plants" (1926).

N.A.Maksimov showed by his experimental data on transpiration intensity in combination with other anatomical and morphological studies that the majority of xerophytes, compared to mesophytes, are characterized not by reduced, but on the contrary by increased transpiration intensity, which is promoted by xeromorphic structure, as a large network of veins and a greater number of stomata contribute to better water removal. Drought tolerance is defined by N.A.Maksimov as the ability to tolerate dehydration, to cope easily after long wilting and with the least damage both to the plant itself and to the yield it brings. Unfortunately, this definition, reflecting the narrow physiological specificity of drought-tolerant plants, has often been used in establishing the survival rate of a wide variety of plants without a clear characterization of their ecological affiliation.

I.M.Vasiliev (1931) makes a number of important conclusions on the basis of ecological studies of plants in the Kara-Kum desert. In his opinion, N.A.Maksimov's concept of wilting is not applicable to all plants of arid territories, but only to those that live in conditions where dry periods are replaced by wet ones.

I.M.Vasiliev (1936), proposes a more detailed classification, namely: division of drought tolerance into biological, physiological and agronomic. Such classification could not solve the existing contradictions on this issue due to the private meaning of the phenomenon drought tolerance.

A great contribution to the systematization of plants of arid habitats from the ecological point of view was made by P.A.Genkel (1939,1946), who on the basis of experimental study and theoretical generalization gives the following ecological-physiological classification of xerophytes: 1) succulents; 2) euxerophytes; 3) hemixerophytes; 4) poikiloxerophytes.

P.A. Genkel suggests that the concept of "xerophyte" should be interpreted only in the ecological sense, without introducing the physiological one, since the ways of xerophytes struggle against drought are very diverse and detailed physiological assessment is possible only for separate groups of xerophytes. P.A.Genkel gives a more precise definition of drought tolerance, emphasizing its physiological specificity. According to Genkel, drought tolerance is "the ability of plants to tolerate dehydration and overheating" (1946).

Many scientists used drought tolerance as a concept of "xerophyte". Thus, the complete identification of the concept of "xerophyte" and "drought tolerance", replacing one term with another, caused great difficulty and was the cause of many inconsistencies and contradictions on this issue.

With the development of structural chemistry and physics, the issues of water regime receive deeper coverage. The experimental works of A.M. Alekseev (1937; 1948; 1957; 1969) and N.A. Gusev (1956; 1959; 1966) should be noted on the studies of water regime of plants in connection with their resistance to drought. These authors established that changes in the structural state of water during drought cause profound disturbances in the structure of protein molecules and in life functions. Under drought conditions, biochemical processes are also changed in plants. In the first phases of water deficit there is an increase in starch breakdown. At dehydration is accompanied by a general increase in the processes of decomposition of high molecular weight carbohydrate compounds to monose. Increase of hydrolytic reactions depresses photosynthetic activity of assimilating plant tissues. Lack of moisture and high temperature leads to disruption of the course of basic life processes. So, drought causes intensification of hydrolytic reactions and delay of synthetic processes (Sisakyan, 1940). Research works of V.N.Zholkevich (1968) are devoted to the study of respiration energetics under conditions of water deficit.

Works on the anatomy of arid plants deserve attention. For example, V.K. Vasilevskaya (1954) found that xerophytes and mesophytes have no sharp differences in leaf structure due to the diversity of their habitats.

An analysis of the history of studies of plant life under drought conditions shows that throughout, plant tolerance and adaptation to drought have been studied mainly in terms of dehydration. The influence of another important factor of drought - temperature - has been studied recently. The first works on this issue were carried out by N.A. Khlebnikova (1937), N.S. Petinov and Y.G. Molotkovsky (1957).

At present, many researchers are working on the problem of physiology and biochemistry of plant heat tolerance. The effect of dehydration and high temperatures on protein and nucleic acid metabolism of plants has been established (Altergot et al., 1966)

Numerous studies on ecological physiology of arid plants have been carried out. A great contribution was made by L.A.Ivanov (1963), B.Z.Huseynov (1952, Y.L.Celniker (1955-1969), M.D.Kushnirenko (1967). These authors worked in different climatic zones, determining various indicators and obtained very important information on ecological peculiarities of plants of arid zones.

4.1. Water regime as an indicator of plant resistance and adaptation to drought.

Studies on water regime are carried out in different directions: changes in water regime indicators are investigated under different soil moisture and in different forest zones (Petinov, 1954; Celniker, 1955;1958 Troitskaya 1959).

Some researchers take the reversibility of dehydration and increase in the amount of bound water as a criterion for determining plant resistance to drought (Kenesarina, 1959; Tereshin, 1960; Kalugin, 1960). Other scientists attach great importance in water exchange to the intensity of transpiration (Blagoveshchensky and Bogracheva, 1955; Krasulin and Pankratova, 1957) and others. More profound works are carried out by scientists to study water regime in connection with metabolism.

Studies on water regime of xerophytes and mesophytes of mountain Tajikistan were conducted by M.I.Matveev (1948, 1953). He determined the following indicators: transpiration intensity, total water content, leaf sucking power and stomata movement in Bukhara almond and walnut. On the basis of the obtained studies, conclusions are drawn about the ecological feasibility of the range of variability of the studied indicators. M.N.Sazonova (1968) states that a characteristic sign of plant adaptation to high temperature conditions is a great lability of transpiration process.

4.2. Terminology used in the study of adaptation of woody plants to drought

As K.A. Akhmatov (1976) testifies, the term "drought tolerance" has been adopted by many scientists as the main term in the study of plant life activity in arid habitats. The ability of plants to grow in arid conditions was assessed mainly by the degree of resistance of cells and organs to dehydration.

Many plants like pistachio, through a variety of adaptations (deep root system, etc.), are virtually immune to significant dehydration.

The current view of dehydration is narrow and the term "drought tolerance" should be applied in a "physiological" sense, and a more succinct expression should be used to characterize the ability of different plants to grow, allowing for the diverse ways in which plants function under drought conditions.

This requirement is met by the words "adaptation" and "adaptation", the meaning of which has begun to expand and clarify in recent years in connection with the mutual penetration of ideas and principles of biology and technology.

V.K.Labutin (1970) points out that although the term "adaptation" has been used in biological literature for a very long time, qualitative shifts in the clarification of its content have occurred in recent years due to the formation and development of the basic ideas of cybernetics. In the author's opinion, adaptation is one of the fundamental and universal concepts of biology, reflecting the dynamic essence of a living system. The property of adaptation makes the organism self-adjusting according to the factors of the external environment in order to find and find the most favorable mode of operation under given conditions. Hence it becomes obvious the diverse ways of settlement and favorable development by plants of various categories of arid lands. In the future, the term "drought tolerance" should be applied only in a limited physiological sense, reflecting the resistance of plant organs to dehydration. In this case, as S.A. Petrov (1972) reports, the use of physiological indicators for diagnostics of comparative drought tolerance will be justified only for closely related species and their varieties.

Thus, it can be recognized that biological adaptation of drought-tolerant plants, including the present pistachio, is primarily conditioned by resistance to dehydration, which is reflected primarily by their water regime.

CHAPTER 5. Peculiarities of the water regime of the true pistachio.

Knowledge of various aspects of plant water exchange expands our understanding of ways of adaptation of plant organisms to environmental conditions. Studies on water regime are conducted in different directions. Some scientists study changes in water regime indicators at different soil moisture (Alekseev and Gusev, 1950 a.b.; Petinov, 1954; Celniker, 1955); others study daily and seasonal changes in water regime of woody plants in different forest zones (Ivanov et al., 1956; Troitskaya, 1959; Celniker, 1958). Some scientists take the reversibility of dehydration and increase in the amount of bound water as a criterion for determining plant resistance to various unfavorable factors (Kenesarina, 1959; Tereshin, 1960). Many scientists attach great importance in water metabolism to the intensity of transpiration (Blagoveschensky and Bogacheva, 1955; Krasulin and Pankratova 1957). In recent years, the work on water regime is more in-depth: joint studies of water regime and metabolism are carried out. Some scientists try to judge the availability of soil moisture by changes in water regime indicators for plants on irrigated and rainfed plots (Kondo, 1946; Petinov, 1954) or for plants of different habitats (Maksimov, 1952; Keller, 1952). Recently, works have appeared (Ivanov et al., 1956; Nasyrov, 1961) on elucidation of interrelations of transpiration and photosynthesis processes in plants.

In Tajikistan, M.I.Matveev (1947,1953) studied water regime of tree species. Continuing these studies, K.P.Rakhmaninova (1962) studied water regime of woody plants growing in natural rainfed habitats on the territory of Varzob Mountain Botanical Station. The task of the research was to find out the peculiarities of adaptation of edifiers of different types of woody vegetation to habitat conditions. During stationary studies of water regime, when daily and seasonal changes in transpiration intensity were studied, an attempt was made to establish the dependence of transpiration on light intensity.

Transpiration has been considered by many scientists as plant resistance to extreme conditions. The works of L.A.Ivanov (1956), Y.L.Celniker and M.I.Markova (1955), M.I.Matveev (1953), K.P.Rakhmanina (1962), K.A.Akhmatov (1970), G.I.Girs

(1963), M.A.Sazanova (1968), K.P.Popov (1971), Vernik R.S, Zakharyants I.L. (1966), touch upon various aspects of studying the water regime of plants. Analysis of the results of these studies shows that these authors have collected a huge amount of factual material on the intensity of transpiration of woody plants under different conditions. The mentioned works emphasize the importance of transpiration as a process characterizing the vital activity of plants, especially in terms of water consumption by both individual plants and entire plantations in extreme climatic conditions. Very often researchers tried to judge by the value of transpiration such a complex phenomenon, which is the resistance of plants to drought period. It is known that the intensity of physiological processes occurring in any plant, including pistachio,

physiological processes occurring in any plant, including pistachio, are related to the conditions of their water supply. Lack of water in soils of semi-desert foothills, where pistachio grows, is not catastrophic for it, thanks to well-developed root system, which assimilates moisture from different soil horizons. Its root system is powerfully developed, has a large absorption capacity. V.N.Gorbunova writes that "no tree species (except almonds) is able to put up with the minimum of moisture that pistachio is satisfied with" (quoted from O.V.Zalensky, 1940).

Assimilatory activity of the pistachio, according to Y.S.Nasyrov (1961), is characterized by a long vegetation duration. Y.S.Nasyrov analyzing the dependence of photosynthesis of pistachio on external and internal factors, notes that the most favorable for assimilation activity air temperatures lie within 26-36^{0}C, and illumination from 30-80-90 thousand lux. Maximum photosynthesis intensity in pistachio (32-38mg/dm^{2}per hour) was found at temperature 31-32^{0}C and illumination 60-64 thousand lux. This indicates, as the author testifies, the exceptional stability and adaptive ability of the assimilative apparatus of pistachio to the action of high temperature and illumination. Photosynthetic apparatus of pistachio is more resistant to dehydration.

The true pistachio is a xeromorphic plant. Thanks to its well-developed root system, it provides itself with the necessary moisture, so even in the dry summer period, the daily dynamics of photosynthesis remains uniform. In spring, at low ambient temperatures, photosynthesis in pistachio has low intensity. In June, as the temperature rises, pistachio has the highest assimilation activity. With the establishment of hot weather (July, August) in summer, photosynthesis slightly decreases, but remains at almost the same, relatively high level until the end of the growing season. In the fall, with the aging of leaves, their assimilative capacity weakens. The daily course of photosynthesis in pistachio throughout the growing season is uniform. The maximum intensity of the process is often observed in the midday hours. At noon, despite high radiation and air temperature, assimilation does not weaken.

When growing the real pistachio, it should be taken into account that in arid areas, as the long-term experience of regional foresters has shown, pistachio crops grow and develop better in pure plantations, without mixing with other, even relatively drought-resistant species. In the first years of its life pistachio develops a very deep root system. If in the first year the above-ground part grows 9-10cm, the root deepens to 90-110cm, sometimes up to 150cm downwards, and lateral roots also develop. For example, in seven years its lateral roots develop up to 4m or more. The area of root feeding according to O.V.Zalensky (1940) is $144m^2$ in pistachio. V.I.Zapryagaeva (1964) found that 50-70 years old pistachio has a projection of horizontal roots 15 times exceeding 25 projection of the tree crown. In the author's opinion, this is one of the reasons for the rarity of pistachio in nature (50-70 trees per 1 ha.). According to O.V.Zalensky, excavations of the root system of pistachio showed that in 23-year-old trees horizontal roots spread to the sides of the trunk for 10-12m. This provides a large area of water and mineral nutrition of the tree, as well as creates favorable light conditions for light-loving pistachio.

The true pistachio, with the help of various adaptations (development of horizontal and vertical root system, etc.) is practically not subjected to significant dehydration. And the importance of water is determined by a number of its properties. It is a solvent and a medium in which movement and metabolism take place. Studies by scientists (V.I.Zapryagaeva and others) have established that the most important life processes in cells occur at a certain water content. Well-adapted species such as Pistacia Vera L, Amugdalus communis, Ulmus pinnato - ramose are characterized by moderate and stable water content in leaves during vegetation.

Stable water content of leaves, due to more active water exchange in xerophilous plants, was established by I.E. Znamensky (1948) and I.M. Matveev (1953). K.A.Akhmatov (1976) believes that in determining plant resistance and adaptation (adaptation) to drought it is not the absolute amount of water in separate periods that is of great importance, but the stability of water content in plant leaves during vegetation, especially during drought, which determines normal vital activity of plants.

High resistance of plants to unfavorable environmental conditions was established by the amount of bound water in plant leaves. According to E.V.Lebedintseva (1926), A.M.Migahit (1945), the amount of bound water in xerophytes is much higher than in mesophytes. D.M.Kushnerenko et al. (1970) consider high content of bound water as one of the indicators of increased drought resistance of some fruit plants.

Studies by K.A.Akhmatov (1976) have established that the stable content of free and bound water in leaves of pistachio is maintained by the activity of a deeply penetrating root system.

With the total water content in leaves during the dry period - from 53.4% in July, 52.3% in August, the amount of free water during this period varies from 22.4% to 22.0%, and, accordingly, bound water - from 31% to 29.4%. According to the authors, this ensures normal (conscious) photosynthetic activity of the assimilation apparatus and reflects the protective orientation of life processes for plant survival in the dry period. At the same time, as K.A. Akhmatov notes, the accounting of water fractional composition, for the purpose of diagnostics and adaptation, as well as adaptation of woody plants to drought, should be carried out depending on the ecological features of plants.

The dependence between water supply conditions and sucking force is the most important adaptation (adaptation) of a plant to one of the necessary factors of its growth - water. V.S. Shardakov (1953) pointed out that under abundant water supply the sucking force is insignificant, and under poor water supply it increases the more, the less saturated plant cells are with water. With decreasing moisture in the soil, the sucking force increases strongly. Thus, according to K.A.A.Akhmatov (1976), for Kyrgyzstan the sucking power of leaves of pistachio from May to August changes as follows: May 28-11atm, July 18 - 21atm, August 25-30 atm, i.e. in arid ecological conditions the sucking power of leaves increases to the end of the vegetation period, and it is also closely dependent on the ecological conditions of plant growth. Increase of sucking power is a characteristic sign of strong water deficit in plants, especially those not adapted (not adapted) to arid arid conditions.

Studies on transpiration of woody plants for Central Asia are covered in the works of M.I. Matveev (1953), K.P. Rakhmanina (1962), K.A. Akhmatov (1962), K.P. Popov (1971), G.M. Chernova and G.S. Olekhnovich (1985). These authors collected a large amount of factual material on the transpiration intensity of woody plants in different growing conditions. It was found that the true pistachio, unlike many xerophytes and mesophytes, is characterized by increased transpiration intensity, which correlates primarily with air temperature, with the lowest value in the morning and evening, and higher in the daytime. That is, pistachio has the highest assimilatory activity with increasing air temperature and light. Cooling of the assimilating surface of plants in hot days (July, August) by means of increased transpiration has a positive effect on the synthetic work of leaves, as the leaf temperature decreases by 3-4^{0}C compared to the environment. The development of a powerful root system of pistachio provides throughout the growing season a high index of transpiration intensity, and this in turn increases assimilation activity in contrast to other tree species and is characterized by a long duration during the growing season, due to wide open stomata during the summer period. Therefore, a great role belongs to the deep root system, which steadily provides underground organs, including leaves, with the necessary amount of water to keep the stomata open.

Plants such as the true pistachio have developed an adaptation in the form of peculiar tap roots with an anatomical structure that allows it to penetrate non-drying layers of soil and from there supply moisture to the leaves.

V.F.Altergot (1963) notes that prolonged overheating of leaves causes deep disturbances in metabolism. Moreover, transpiration acts as the main physiological factor determining leaf temperature. In pistachio and common almond, the temperature gradient between the leaf and the surrounding air during the whole daylight period is negative, i.e. the leaf temperature is lower than the air temperature. The increased level of transpiration, which gives a high thermoregulatory effect, is a factor of paramount importance. S.I.Kokina (1935) and Lange (1959) managed to prove that high transpiration intensity of desert plants allows them to exist in extremely arid conditions. K.P.Popov (1974) emphasized that leaf temperature of most species in July and August exceeds the ambient air temperature. The exceptions are pistachio and almond, which retain high thermoregulatory ability through increased transpiration until the end of vegetation. Cooling of the assimilating surface of plants on hot days by means of increased transpiration has a positive effect on the synthetic work of leaves. This is confirmed by the high level of photosynthesis of pistachio and almond during the whole vegetation season, including hot and dry periods.

CONCLUSIONS.

a) Many scientists accept the reversibility of dehydration as a criterion for revealing plant resistance (adaptation) to various unfavorable factors. High transpiration has been considered by many scientists as a stabilizing factor of plant resistance to extreme conditions. At the same time, assimilation activity in pistachio is characterized by a long vegetation duration.

б) Under arid ecological conditions, leaf sucking power of pistachio increases from the beginning to the end of the growing season from 11atm. (May) to 30-40 atm (August).

c) It has been established in Central Asia that the true pistachio is characterized by increased transpiration intensity, which correlates primarily with air temperature. Long-term overheating of leaves causes deep disturbances in metabolism, with transpiration acting as the main physiological factor stabilizing leaf temperature.

Therefore, to establish the high adaptive capacity of pistachio, which is confirmed by the wide geographical and altitudinal range of natural distribution of this species within Central Asia, in the period 2006-2012, studies were conducted on the prospects and possibility of pistachio cultivation in very harsh natural-climatic conditions characteristic of gravel lands in the Fergana forest region of Uzbekistan. Moreover, elements of water regime were studied not only in young plants, but also in varieties first released on plantations established on pebbles.

CHAPTER 6. Studies on adaptation of pistachio plants grown on gravelly lands in the Fergana Valley.

As noted above, Central Asia is the center of origin of a valuable nut-bearing species - the true pistachio. In the Fergana Valley it grows on the southern slopes in the foothills of the Fergana, Chatkal, Turkestan, Alai ranges, which are the northern geographical coordinate of the range of this species, as well as in the foothills and lowlands of the Gissar and Babatagh ranges (the southern geographical border of the range of this species in Uzbekistan), where pistachio forms a belt of pure "pistachio sparse forests" in a wide range of absolute altitudes from 500 to 1800m above sea level, occupying mainly dry, well-warmed slopes on light sandy loamy soils of gray soil type. In addition, both on the northern and southern boundaries of the range, pistachios are found on steep stony slopes. This predetermines the possibilities and prospects of cultivation of this species on gravel lands of valley parts of the foothills of the Fergana Valley.

Since ancient times, pistachio in Uzbekistan was cultivated on homestead plots, mazars and burial mounds were planted with it. According to V.I.Massalsky (1913), B.A.Fedchenko, A.I.Krishtofovich (1914), at the beginning of the 18th century, thickets of fruit species, including pistachio, not only covered mountains in the Fergana Valley, but also descended to cultural oases.

Now there are practically no pistachios there. But in the recent past, the Fergana Valley was a "truly pistachio country".

Pistachio cultivation at the state level in Uzbekistan began in the post-war period, when the Union Ministries of Forestry were organized. This included forestry production organizations - leskhozes. Pistachio cultivation was established mainly for the purpose of reclamation of rainfed foothills, and on the territory of the present-day Saraykurgan forestry center, starting from 1949 - for the purpose of afforestation of the territory around the Kattakurgan reservoir.

In addition, it should be emphasized that it is in forest cultures, created by sowing seeds from different maternal individuals, concentrated valuable gene pool of diverse forms of pistachio, where by now selected more than a dozen of large-fruited forms of pistachio, whose fruit with a well opening bones and high yield of kernel.

Figure 3 Pistachios of Uzbekistan

By its biological peculiarity, pistachio prefers warm, well-warmed slopes with light medium loamy sierozem. Therefore, the attempt to introduce pistachio into culture in almost new forest conditions on very fertility-poor gravels of the river outflow cone, when the upper soil horizons are saturated with coarse rubble and pebbles, has been carried out for the first time since 2006.

At the same time, gravels of the drainage cone of individual rivers occupy huge spaces along many foothills of mountain systems: from the Dzungarian Alatau in the north to the Gissar Range in the south. It is recognized that the development of new lands, previously considered "abandoned", based on their irrigation and reclamation, can solve an important problem of increasing their soil fertility and, accordingly, will open prospects for their use for agricultural and fruit crops.

The objects of stationary research, where pistachio cultivation has been started, are located on the lands of Kokand forestry, located in the valley part of Turkestan ridge, as well as on the lands of Kokand forestry experimental station, in the valley part of the foothills of Alai ridge, are part of Kokand group of areas with intense wind activity, where undeveloped pebble lands are subjected to active wind erosion. Strong winds up to 25-26 m/sec in March and April blow out fine earth from the surface layers of pebble, exposing pebbles and rubble, reducing the fertility of already poor pebble lands.

6.1. Soil-climatic characteristics of the research areas.

The true pistachio is characterized by good adaptation to different growing conditions, withstanding both extremely dry climate at lower altitudes and relatively low winter temperatures at higher altitudes. On some ridges, such as Babatagsky, Gissar and others, pistachio grows as single trees and groups of trees on steep stony and rubbly soils.

At the same time, by its biological peculiarity, pistachio prefers warm, well-warmed slopes with light medium loamy sierozem.

Therefore, the attempt to introduce pistachio into culture in practically new forest conditions on very fertility-poor gravels of the river outflow cone, when the upper soil horizons are saturated with coarse rubble and pebbles, was carried out for the first time.

At the same time, gravels of the drainage cone of individual rivers occupy vast areas along many foothills of mountain systems: from the Dzungarian Alatau in the north to the Gissar Range in the south. It is recognized that development of new lands, previously considered as "abandoned", on the basis of their irrigation and reclamation can solve an important problem of increasing their soil fertility and, accordingly, will open prospects for their use for agricultural and fruit crops.

As T.A. Zheltikova (1968) testifies, long-term experience of SredazNIILKh scientists has shown that when growing on gravels under irrigation conditions of field-protective forest strips, forest and fruit species, they fulfill the role of an active ameliorating factor, improving microclimate, increasing soil fertility due to accumulation of organic matter. It has been revealed that changes occurring in the soil under the forest or garden canopy are incomparably faster than under the cultivation of annual agricultural plants. These changes are expressed in the increase in the thickness of humus horizons and humus content in them, improvement of structure and input-physical properties of the soil, and are manifested already in 6-7 years after planting a forest or an orchard.

The undeveloped gravel lands of the study area are very poor in soil nutrition elements, especially organic matter. This is evidenced by very scarce natural herbaceous vegetation on these lands, which in hot dry climate, of course, does not contribute to the accumulation of organic matter in the soil. In order to obtain objective data, the study of soil environment and the content of basic elements of soil nutrition (nitrogen, phosphorus) was carried out at the sites of experiments.

Data of the nearby station "Kokand" were used to characterize the climate. The climate of the district is characterized by relatively cool winter and long hot summer. The coldest months are January, February, November and December. With the average monthly minimum temperature in January and December - 5.O^{0} C, the absolute maximum in these months can fall to - 8.2 and 10.O^{0} C. Spring months are characterized by gradual increase of positive temperatures and already in April it reaches 16.7^{0} with the maximum value in some days 32.4^{0} (Table 4). The last spring frosts (up to - 4.0$^{0)}$ occur at the end of March, and sometimes even in April. In summer months air temperatures are relatively stable, with monthly averages in June 27.3^{0}, July 28.7$^{(0)}$, August 25.4(0), with an absolute maximum from 37.0$^{(0)}$in August to 41.2^{0} in July.

With a mean annual precipitation of 212.5 mm, its maximum value is observed during spring showers in April (66 mm) and May (61 mm) and practical absence from June to September.

As already mentioned, the study area is characterized by intense wind activity. Here the winds, mainly of south-western

directions, practically observed throughout the entire annual cycle, with variations from 13 to 26 m/sec.

In general, by temperature gradients and annual precipitation, the study area can be attributed to the zone of very arid arid foothills, characteristic of the zone of semi-desert foothills of many ridges in Uzbekistan, including semi-desert foothills of the Nurata Ridge on the territory of the Saraikurgan forestry, where pistachio crops have already been grown around the Kattakurgan reservoir on an area of more than 3 thousand ha.

During the years of research (2006-2014), the weather situation was generally close to the mean annual characteristics. The exception was lower average daily air temperatures in March and April in 2009 (by 3-5^{0}C) and, in general, in these months precipitation was 15-20 mm more than the mean annual data.

Table 4

Mean annual data of some elements of climate in the area of research (based on the weather station "Kokand").

Months	Air temperature $_0$C				Precipitation, mm	Relative air humidity, %	Wind force, m/sec.
	cf.	Maxim.	minimum.	abs. min			
January	3,2	11,0	-5,0	-8,0	11,7	84	13
February	2,6	15,9	-4,1	-7,5	15,7	80	16
March	13,2	24,0	0,0	-4,0	28,4	68	26
April	16,7	32,4	0,5	-2,5	66	60	25
May	20,8	33,8	9,0	5,5	61	61	17
June	27,3	40,5	17,0	14,0	0,9	49	22
July	28,7	41,2	15,0	12,0	0,0	46	14
August	25,4	37,0	8,0	8,0	4,7	59	19
September	23,3	36,5	6,0	6,0	2,6	59	14
October	15,8	31,0	2,0	0,0	8,7	66	19
November	8,6	20,0	-3,7	-6,0	10,9	78	16

December	3,5	11,0	-5,0	-10,0	11,6	81	14
Average annual	15,7					66,3	18
Total precipitation for the year					222,2		

6.2. Study of soil environment and content of basic elements of soil nutrition (nitrogen, phosphorus) at the sites of experiments.

In order to study the soil environment at the sites of pistachio cultivation, a soil transect was laid in 2009 on gravels not previously used for agricultural crops.

Here, there is a crushed stone and pebble backfill on the surface, which was formed due to blowing of fine-grained gravel from the surface of the colmatation horizon layers by strong winds. The upper (0-20 cm) colmatation horizon of pale gray loam contains about 15-20% pebbles (medium-skeleton fine-grained, compacted).

The 20-25 cm thick transitional horizon still contains 30% fine-grained sandy loam, but 70% is composed of skeletal inclusions (pebbles, rubble), so it should be defined as a strong fine-grained-skeletal horizon.

The amount of humus in soils of virgin origin. The 0-60 cm horizon contains 0.12% of humus (Table 5).

Table 5 Agrochemical characterization of soils of experimental plots.

Horizon	In % of soil weight			Mg/kg soil		
	Humus	total nitrogen	total phosphorus	N-NO3	N-NH4	P2O5
0-20	0,12	0,05	0,20	6,75	22,1	28,0
20-40	0,12	0,05	0,14	6,70	24,6	28,0
40-60	0,12	0,05	0,15	6,75	25,3	24,0
60-80	0,86	0,10	0,20	4,50	24,6	34,0
80-100	0,55	0,05	0,20	5,62	24,6	24,0

It should be noted that humus content on the experimental objects is typical for this type of virgin pebble lands in the Fergana Valley, and it can be categorized as weak-skeleton-small-gravel.

According to the amount of total nitrogen, soils contain 0.05-0.08% nitrogen. The results of determination of mobile forms of nitrogen (NO_3 and NH_4) in initial samples show that the amount of nitrate nitrogen in soil does not exceed 5.62-6.75 mg per kg of soil. The amount of nitrate is distributed rather uniformly over the soil profile. The amount of ammonia form of nitrogen significantly exceeds the content of nitrate nitrogen and is characterized by its more uniform distribution on the meter thickness. The content of ammonia nitrogen in the 0-100 cm layer varies within 22.1-27.2 mg/kg of soil. Also more uniform distribution over the soil profile is noted for the content of assimilable phosphates. In the meter layer of soil the content of mobile phosphorus does not exceed 24,0-38,0 mg/kg of soil. According to the scale of nitrogen and phosphorus availability developed by the former SOYUZNIIKhI, soils belong to the category of extremely low availability.

Soils are practically non-saline (Table 6). Dense residue, according to the data of water extract analysis, does not exceed 0.5%, and the value of dense residue remains almost unchanged along the whole profile of meter layer. Chlorine content in the composition of water-soluble soil salts at both sites is insignificant 0.01-0.007%. The content of "SO4" ions in the soil does not exceed 0.24-0.25%.

Table 6

Results of the analysis of the aqueous extract.

Horizon, cm	In % of soil weight							
	Dense residue	NCOz^1	CI^1	SO4	Ca$^+$	Mg$^+$	Na$^+$	Sum of salts
0-20	0,460	2,20	0,30	5,12	5,50	1,50	0,62	0,432
		0,033	0,011	0,246	0,110	0,018	0,014	
20-40	0,455	2,25	0,15	4,99	5,25	2,00	0,19	0,412

		0,034	0,005	0,240	0,105	0,024	0,004	
40-60	0,445	2,0	0,20	4,99	5,25	175	0,39	
		0,030	0,007	0,240	0,105	0,021	0,009	0,407
60-80	0,435	1,85	0,25	4,99	4,50	175	0,84	
		0,028	0,009	0,240	0,090	0,021	0,019	
80-100	0,470	1,80	0,25	5,21	4,75		2,51	0,438
		0,027	0,009	0,250	0,095		0,057	

The content of carbonates ($CaCO_3$) throughout the soil profile does not exceed 6-7%; the 0-40 cm layer contains 0.18% $CaCO_4$ (Table 7).

Table 7

Individual definitions, %.

Horizon, cm	Specific gravity of soil, g/cm^3	Gypsum ($CaSO_4$)	Carbonates ($CaCO_3$)
0-20	2,78	0,14	6,60
20-40	2,76	0,15	6,38
40-60	2,63	0,18	7,04
60-80	2,69	0,15	6,71
80-100	2,64	0,18	6,60

The results of the soil analysis data from the experimental plots at the two sites lead to the following conclusions:

1. The granulometric composition of soils is dusty-clayey with transition to gravels. They are low-structured and difficult to be cultivated.

2. In terms of humus, gross nitrogen and phosphorus content the soils are not characteristic and can be referred to pebble soils.

3. According to the content of mobile forms, especially nitrates and phosphates, soils on zakolmatized gravels belong to the group of unsecured and should be well responsive to fertilizer application, which was practically confirmed during the study of mineral fertilizers application on growth and development of young pistachio plants

on plantations established in 2006-2009 on the territories of Kokand forestry and Kokand Forest Experimental Station.

CHAPTER 7. Study of the effectiveness of the application of biologically active growth substances in a plantation created by planting seedlings grown in containers with closed root system (CRS).

The solution of this problem is based on the development of pebble lands for pistachio crops not by the traditional method of sowing seeds on a permanent place, but the use of planting material with a closed root system (RCS). The basis of our proposed new more effective technology is the method of growing pistachio seedlings in containers of small volume (0.1251.) on the type of "seedlings".

Containers are prepared from polyethylene film. The size of the bag is 5x25cm. Small holes are made in the bottom of the container, and then they are filled with soil substrate consisting of 5 parts of soil and 1 part of over-fermented manure. To avoid infection of seedlings by fungal diseases, the soil is taken from areas not previously used for nightshade crops. If possible, the prepared substrate is fumigated with methyl bromide. Containers are filled with substrate and placed in boxes of 100 pieces each. Seeds are sown into containers in the first decade of February. Stratified seeds are used for sowing.

7.1. Advantages of the method

- Provides high rooting rate and good plant growth.
- Extends planting dates by 2 months.
- Reduces seed consumption.
- The lightweight containers (20 kg box weight) can be easily transported to the planting site.
- It takes 2 boxes of planting material to develop 1 hectare of land.

When creating pistachio plantation crops on gravel lands in the territory of Kokand forestry by sowing pistachio seeds on a permanent place, a positive result on seedling survival rate (75-80%) was obtained. However, in order to obtain guaranteed germination of at least one or two seedlings, up to 8-10 pcs of stratified seeds were sown on the seedbed.

Experiments on the development of technological methods of growing pistachio plantations on zakolmatized gravels were laid in 2 variants - plantation establishment by planting seedlings grown in containers of small volume (5x25 cm) and sowing seeds on a permanent place. The planting scheme was 6x8 m. Plantations were established on the area of 5 ha on the territory of Kokand forestry and on the area of 4 ha on the stationary site of Kokand Forest Experimental Station (valley part of the foothills of the Alai Range). Experiments to study the effectiveness of application of growth substances were laid on the stationary site of the Kokand Forestry Experimental Station.

Containers (polyethylene bags with small holes on the sides at the bottom for ventilation of the root system) were filled with substrate consisting of three parts of soil and one part of over-fermented manure. Ready (stratified) seeds were sown in containers in the third decade of February - first decade of March. Sowing depth was 3 cm, one seed in each container. When sowing pistachio seeds directly into the ground, sowing was carried out in the first decade of March at a depth of 5-7 cm, 7 seeds per hole (line).

To stimulate germination and increase seed germination from humic acid salts, the preparation "rooting agent" (RA) - humaton of 60 % concentration was applied:

1 .UK-2 (2g. per 10 liters of water);

2 .UK-4 (4g. per 10 liters of water);

3 .UK-6 (6g. per 10 liters of water);

4 . Soak in clean water (control).

Stratified pistachio seeds were treated by soaking them in specified solutions for 24 hours. The control seeds were kept for the same number of hours in ordinary drinking water. Seeds of Orzu variety were used in the experiment and in the control.

The results of the first year of research showed that when sowing pistachio seeds soaked in the given solutions of humatone, their germination in containers varied from

70 to 95%, with a maximum value of 95% in the variant UK-2 and a minimum of 70% in the control. Higher doses of this preparation (4 and 6 g per 10 liters of water) reduced seed germination by 10-15%.

Figure 4: Containers with closed root system planting materials (RCPP)

It also follows from the data of Table 8 that pre-sowing treatment of seeds with growth substances stimulates the intensity of development of above-ground parts of plants. At the average height of seedlings in containers before planting PMZK in the ground amounted to 12.4^{+} - 0.03 cm, at the end of the first year this indicator, due to the growth of plants in height, amounted to 15.7^{+} - 0.03 cm, with a maximum value of 17.3^{+} - 0.03 cm in the variant UK-2. Accordingly, on the control (soaking seeds in clean water), the average height of seedlings at the end of the first year of growth was 12.8^{+} - 0.04 cm, i.e., the differences in growth intensity are reliably provable ($t<3$).

At the same time, when pistachio seeds were sown directly into the ground, seedling survival rates ranged from 65-75%, with an average seedling height, when recorded at the end of the growing season (October) of 7.9+. 0.03 cm

Table 8

Effect of application of growth agents on germination of pistachio seeds and growth of container-grown seedlings.

Experiment variant	Number of sown seeds, pcs	Germination, % of the number of sown seeds	Containers (PMPC)	
			Average height before planting	Average height at the end of the first year
1. UK-2	100	95	12,8+- 0,03	17,4+- 0,03
2. UK-4	100	80	12,3+. 0,02	16,1+- 0,03
3. UK-6	100	75	12,1+- 0,02	14,3+- 0,03
Control	100	70	10,5+- 0,03	12,8+- 0,04
Seed sowing	100	65	-	7,9+- 0,03

Thus, the preliminary results of using small volume containers for planting pistachio plantations with the use of growth stimulants, obtained already in the first year after planting on pebble lands, testify to the promising potential of this method for growing this valuable species in very harsh forest conditions.

7.2 Study of the effectiveness of the use of film-forming water-retaining structurants (structurants) on growth, state and water regime of young pistachio plants.

In experiments on zakolmatized gravels under this project structurant "ameliorant" was applied in the first decade of April 2009 to the seedbeds of planted seedlings of pistachio PMZK. The treatment area was 0.5 square meters. The experiment was laid in 3.x variants: 1. 0.2% concentration of aqueous solution; 2. 0.4% concentration of aqueous solution; 3. control (without structurant application). Each variant and control had 75 plants.

It is known that the intensity of physiological processes occurring in any plant, including pistachio, is related to the peculiarities of its water regime, including transpiration intensity, leaf wetness, water deficit and leaf sucking power. Previous studies conducted in pistachio cultivation in optimal forest conditions for this species, in rainfed foothills on soils of serozem type, as well as in pistachio cultivation in very harsh arid foothills, including on gravels in Fergana oblast (data of observations on water regime in the period 2006-2009) have established that favorable water regime in pistachio cultivation in very harsh arid foothills, including on gravels in Fergana oblast.It was found that favorable conditions of water availability cause more intensive physiological processes in plants (plant growth, obliquity, as well as higher indicators of transpiration intensity, etc.).

As the observations on transpiration intensity of pistachio PMPC in the experiments with application of structurant showed, already in the first year after planting, young plants responded to the improvement of moisture supply conditions by increasing their daily and seasonal transpiration intensity values. As shown in Table 9, the average seasonal values of transpiration intensity of plants in the experiment with the application of "meliorant", both at 0.2% and 0.4% concentrations, exceed those of the control by 16-20%.

Table 9

Transpiration intensity (mg/g/hour) in annual seedlings of pistachio on the variants of the experiment with the application of structurant "meliorant".

Months	Hours of observation					
	9-11	11-13	13-15	15-17	17-19	Average daily
0.2% concentration						
May	833	1300	2358	1212	898	1320,0
June	842	1241	2410	1398	975	1373,2
July	1155	1300	2463	2410	1398	1745,2
August	1520	2010	2583	2704	1250	2013,4
September	723	1810	2010	2529	1120	1646,4
Midseason						1619,6
0.4% concentration						

May	1403	1722	2053	2130	1330	1727,6
June	1633	1929	2202	2366	1407	1907,4
July	1947	2572	3157	3255	1557	2497,6
August	1384	2481	3076	3250	1227	2283,6
September	1330	2229	2553	2360	1115	1917,4
Midseason						2066,7
Control I	without the introduction of a structurant)					
May	726	840	1129	1347	870	982,4
June	851	1027	1720	1960	916	1293,8
July	1183	1570	1860	2640	1127	1676,0
August	1012	1520	1780	2425	1095	1566,4
September	920	1416	1615	2270	910	1426,2
Midseason						1388,9

At the same time, higher, compared to the control, indicators of transpiration intensity in pistachio seedlings with the application of structurant caused an increase in the deficit of leaf water saturation, which was replenished at night in the absence of evaporation during this period. Thus, if on the experiment variants with the application of "meliorant" water deficit at the solution concentration of 0.2% amounted to 6.72%, at the solution concentration of 0.4% amounted to 8.24%, on the control in the seasonal dynamics it did not exceed 5.5%.

Leaf humidity of annual pistachio seedlings, both in the experiment and in the control, fluctuated within 55.9-60.6%, not falling below 50% of the level of water content per leaf wet weight. This in turn indicates not only the positive response of young pistachio plants to growing conditions, but also in general to favorable forest conditions in which plantation crops of this exceptionally drought-resistant species are grown with the application of structurant.

Consequently, the use of film-protecting preparations improves the conditions of moisture availability of pistachio plants, providing more favorable conditions for their growth and condition. From the data of taxation indices of annual plants obtained at the end of the vegetation period, a more intensive growth in height and diameter can be traced in the experiment variants with the application of structurant. On average, the indices of their height, stem diameter and current linear growth exceeded the control ones by 14-16% at the experiment accuracy t>3 (Table 10).

Table 10

Growth of annual seedlings of pistachio on variants of the experiment with application of structurant "meliorant" (average indicators).

Experiment variant	Annual seedlings (PMPC) M^{+} - m		
	Height, cm	Diameter at the root neck, mm	Current linear growth in height, cm
0,2%	17.4^{+} -0.03	3,9	4.9^{+} -0.01
0,4%	21.2^{+} -0.03	4,3	8.7^{+} -0.02
Control	12.8^{+} -0.03	3,3	3.8^{+} -0.01

Thus, the preliminary results on the development of effective methods of pistachio plantations establishment on zakolmatized gravels allowed to conclude that the areas of the valley part of the foothills of the Alai and Turkistan ranges by temperature gradients and annual precipitation sum can be attributed to the zone of very arid foothills, on soils of sulfur-soil type, characteristic of the zone of rainfed semi-desert foothills, mainly promising for cultivation of drought-resistant pistachio.

However, the development of pebble lands even for cultivation of one of the most drought-resistant tree species and, in general, relatively not demanding to soil fertility, is associated with its cultivation on low-structured, difficult to cultivate pebble gravels.

Therefore, the development of a set of more accelerated technological methods that ensure the efficiency of creating industrial plantations of pistachio by using seedlings grown in containers with a closed root system, with the use of growth substances and film-protecting structurants, will make it possible to involve into agricultural turnover huge desert "abandoned" pebble lands and to obtain valuable pistachio products in the future.

Application of small volume pistachio PMPC of the present small volume by the "seedling" type allows not only to reduce seed costs 5-6 times, but also to significantly lengthen the planting period (up to May inclusive). The use of biologically active growth substances and structurants in the cultivation of PMPC in complex allows to develop more effectively and efficiently the uncomfortable categories of pebble lands in the Fergana Valley for pistachio plantations, the area of which exceeds 400 thousand ha.

CHAPTER 8. Biological adaptation of pistachio plants through the study of water regime elements.

It should be emphasized that water regime of plants - intensity of water release, leaf humidity, deficit of leaf moisture saturation, is an important physiological process reflecting the reaction of plants to the conditions of daily and seasonal dynamics of these processes in young pistachio plants in new growing conditions, namely, zakolmatized gravels of the outcrop cone with a given irrigation regime (from April 12 to September) and was the purpose of research to identify the adaptation of pistachio plants under cultivation in harsh lakes (from April 12 to September).

The works of a number of researchers: Rakhmanina K.P. (1962); Popov K.P. (1979) Chernova G.M., Olekhnovich G.S. (1985) established that the true pistachio, unlike many xerophytes, as well as mesophytes, is characterized by a very high intensity of transpiration (IT), which correlates primarily with air temperature, with the lowest value in the morning and evening hours and a higher value in the daytime hours. In seasonal dynamics also the highest transpiration intensity of this breed in summer hot months. That is, with increasing air temperature, pistachio has the highest assimilation activity. At the same time, if the deficit of leaf moisture saturation correlates with the presence of available moisture in the soil, then leaf moisture in pistachio, under favorable growing conditions, as a rule, is a constant value, slightly increasing in spring, decreased by the end of vegetation, but not falling below 50% of their water content.

Thus, the study of the complex of physiological processes in young pistachio plants in relatively harsh forest conditions of their cultivation will allow us to assess the possibility of development of huge vacant areas of the valley part of the foothills of the Alai and Turkestan ranges for plantations of this valuable nut crop.

The study of water regime dynamics was carried out in 2008 on 3-year and in 2012 on 5-year plantation, established in the areas of research with the scheme of planting 6x8 m (208 plants per 1 ha). 10-15 plants typical in their development were selected for the study. Repetition of the experiment was 4-fold.

The results of IT study in daily and seasonal dynamics are given in Table 11. Transpiration was counted 2 times a month from May to September inclusive, according to the method of quick weighing by L.A. Ivanov (1963) in leaves of the illuminated part of the crown, followed by their 3-minute exposure and re-weighing.

As can be seen from the data presented in Tables 11-16, pistachio in Z-i-5 years of age under new rather extreme growing conditions retains its inherent feature of more intensive water yield during the daytime (from 13 to 15 hours) in the summer (July) hotter period of the year. At the same time, if in July its average values were 2692 and 3056 mg/g/hour, in September - 1069 mg/g/hour, which generally agrees with the data of K.P.Popov (25), studying the intensity of transpiration in adult pistachio in the Aruktau Ridge (Southern Tajikistan). On average, three-year-old seedlings use 159.1 m^3/ha of water for transpiration, or 15.9 mm (Table 11).

The study of moisture and leaf moisture deficit in daily and seasonal dynamics showed that pistachio plants did not show leaf moisture deficit from spring, until the end of the growing season (Table 11;

Table 11.

Mean transpiration rates in 3-year-old pistachio plants mg/g/hour.

Months	Hours of the day						Average
	7-9	9-11	11-13	13-15	15-17	17-19	
May	625,75	1112,2	1911,5	2170,5	1308,5	916	1340,8
June	1374	2510,7	3070,0	3777,8	1875,8	1116,0	2254,0
July	1749	2591,5	3393,5	3852,0	2842,8	1724,5	2692,2
August	643,5	1079,2	1888,75	2217,5	1383,75	787,5	1333,4
September	627,5	943,25	1492,5	1755	1052	545	1069,2

Table 12.

Water deficit (%) of 3-year-old true pistachio.

Months	Hours of the day						Average
	7-9	9-11	11-13	13-15	15-17	17-19	

May	9,2	9,4	10,2	10,6	12,0	13,3	10,7
June	9,4	9,8	11,2	12,7	13,3	14,8	11,8
July	9,7	10,3	12,4	13,6	14,9	16,4	12,8
August	8,8	9,4	11,2	13,0	13,3	14,1	11,5
September	5,7	7,2	9,9	10,4	11,2	12,8	9,5

Table 13.

Water content in leaves of 3-year-old true pistachio leaves (% per crude mass).

Months	Hours of the day						Average
	7-9	9-11	11-13	13-15	15-17	17-19	
May	63,9	61,8	58,9	58,8	60,8	62,1	61,0
June	63,6	61,0	60,6	56,3	58,0	62,0	60,0
July	60,6	58,0	57,0	58,6	59,1	60,3	58,1
August	60,4	59,8	57,4	55,4	60,8	61,2	59,2
September	60,0	58,0	56,8	55,9	58,0	63,8	58,7

Table 14.

Average transpiration rates in 5-year-old plants of true pistachio (mg/g/hour).

Months	Hours of the day						Average
	7-9	9-11	11-13	13-15	15-17	17-19	
May	707	1213	2002	2780	1670	1060	1605
June	1620	2570	3200	3880	2600	1270	2523
July	1840	2920	3870	4210	3540	1960	3056
August	1210	2440	2880	3320	2780	1850	2413
September	770	1380	1870	2130	1350	880	1396

Table 15.

Water deficit (%) of 5-year-old true pistachio.

Months	Hours of the day						Average
	7-9	9-11	11-13	13-15	15-17	17-19	
May	10,3	10,8	11,4	12,6	12,8	13,2	11,8

June	10,7	11,2	11,8	13,2	14,1	14,9	12,6
July	10,9	11,8	12,7	14,8	15,6	16,8	13,7
August	9,4	10,5	12,3	13,5	14,7	16,2	12,7
September	6,7	10,2	H,7	11,9	12,4	13,6	11,1

Daily dynamics of leaf water deficit, both in three-year-old and five-year-old seedlings, is characterized by the lowest values in the morning hours (7-9 h) and amounts to 5.7 in the first case and 6.7% in the second. This is due to comparatively lower air temperatures during these hours. And the daily maximum of water deficit falls on the hotter summer period and, as a rule, is observed in the evening hours (17-19 h), when there is a sharp decrease in water consumption by pistachio leaves for transpiration.

These values are 16.2 and 16.8%, respectively.

Table 16

Water content in leaves of 5-year-old true pistachio leaves (% per crude mass).

Months	Hours of the day						Average
	7-9	9-11	11-	13-	15-	17-	
May	65,4	64,5	60,2	59,6	59,9	63,2	62,2
June	65,2	63,4	61,4	57,2	58,6	63,0	61,4
July	61,3	60,8	58,7	56,4	57,8	60,9	59,3
August	61,2	60,0	58,4	55,8	58,4	61,4	59,2
September	60,8	59,2	58,5	55,2	57,7	62,3	58,9

In seasonal dynamics, the highest values of water deficit occur in the hottest July and are 12.8% in 3-year-old pistachio and 16.8% in 5-year-old pistachio compared to, 9.5 and 11.1% in September.

The amplitude of daily fluctuations of water content in pistachio leaves in the study area decreases with increasing water deficit. The daily maximum of leaf water content is observed in the morning hours (7-9 h) and ranges from 60.0 to 63.9% in 3-year-old seedlings and from 60.8 to 65.4% in 5-year-old seedlings. The minimum of these indices are recorded in the afternoon hours (13-15 h); i.e., they occur during the hottest period of the day, when there is a sharp increase in transpiration intensity and make up

55.9-58.8% and 55.2-59.6%.

Many scientists have considered transpiration intensity in the study of plant resistance to extreme conditions. Very often researchers tried to judge by the transpiration value such a complex phenomenon as resistance (adaptive capacity) of plants to dry period (M.I. Matveev, 1953; K.P. Popov 1971; K.P. Rakhmanina 1962; K.A. Akhmatov 1970).

It has been established in the literature that well-adapted woody plants are characterized by rapid growth and deep penetration of the root system, while poorly adapted ones are characterized by slow growth and superficial root spread. Strong root growth is observed in more adapted species. In pistachio, two tiers are noted in the arrangement of skeletal roots. The first one consists of roots of horizontal direction. The more powerful second tier in pistachio starts from 1.0 m depth and has widely branched roots (K.P. Popov 1971).

The pistachio is a xeromorphic plant; due to its strongly developed root system, it provides itself with necessary moisture, so even in dry summer period the daily dynamics of photosynthesis remains uniform. Assimilation activity of pistachio, according to the data of Y.S. Nasyrov (1961) is characterized by a long daily and vegetation duration, due to good water exchange. Besides, photosynthesis of pistachio depends on external and internal factors, it is proved that air temperature is favorable for assimilation activity within 26-36°C, and illumination from 30 to 80-90 thousand lux. The maximum intensity of photosynthesis Yu.

Nasyrov (1961) notes in pistachio (32-38 mg/dm^2 per hour) was found at temperature 31-32 °C and illumination 60-64 thousand lux. This indicates the exceptional stability and adaptive capacity of the assimilation apparatus of pistachio to the action of high temperature and illumination. The photosynthetic apparatus of pistachio is more resistant to dehydration.

CHAPTER 9. Study of daily and seasonal dynamics of water regime of varieties to establish the response of pistachio to new growing conditions

As observations on water regime of young (3-5 years old) pistachio plants grown on pebble lands on the territories of Kokand forestry (site "Shoberdi" in the foothills of Turkestan ridge) and Kokand LOS (foothills of Alai ridge), conducted in the period 20092012, showed that the pistachio showed high adaptation to harsh forest conditions, characteristic of the zone of formation of pebbles in the Fergana Valley. Practically, in the absence of fertile soil horizon, high summer air temperatures, low relative humidity, intensive wind activity, pistachio in new conditions retained inherent in this species high daily and seasonal transpiration intensity, almost constant water content in leaves and relatively insignificant water saturation deficit in leaves. The same regularity was revealed in 3 pistachio varieties, first released at plantation establishment in these forest conditions.

In order to establish the response of pistachio varieties vegetatively propagated on young 6-year old rootstocks on the plantation in Kokand forestry, the study of water regime elements was included in the complex of observations (Table 17;18;19;).

Table 17

Transpiration intensity of 6-year old pistachio of true pistachio in seasonal dynamics depending on the variety

Transpiration intensity, mg/g/hour					
May	June	July	August	September	Seasonal average
Mountain Pearl					
1988.2	2348.2	2917.3	2524.3	2248.2	2405.2
The Orzu variety					
1728.2	1953.4	2787.8	2629.3	2275.6	2274.8
Albina variety					

1988.4	2009.5	2568.3	2045.6	1568.5	2036.1
Control (non-grafted plants)					
2521.4	2828.2	3112.5	2254.8	1826.8	2508.7

As can be seen from the data presented in Table 17, transpiration intensity of both varieties and non-grafted seedlings increases from spring to summer, reaching its maximum in July-August. According to I.N. Beydeman (1960), this rhythm of transpiration intensity, which increases by summer and decreases by fall, is peculiar to plants constantly supplied with moisture. According to I.N. Beydeman's classification of three rhythms of transpiration intensity, in Fergana Valley conditions on gravelly lands, pistachio can be referred to the first type of seasonal rhythm, i.e. transpiration intensity increases until mid-summer, falling by fall.

Real pistachio is characterized by a relatively high and stable water content in leaves, which, as a rule, does not fall below the 50% mark. The same pattern was observed in both varieties and unvaccinated plants, when leaf moisture varied from 57.5% to 59.4% in varieties and from 52.3 to 58.4% in control unvaccinated plants.(Table 18).

Table 18

Moisture content of leaves of 6-year old pistachio depending on variety

Leaf moisture in % per raw weight					
May	June	July	August	September	Seasonal average
Mountain Pearl					
59.2	56.7	57.2	59.3	62.4	58.96
The Orzu variety					
59.8	56.1	54.8	58.5	61.4	58.12
Albina variety					
57.9	54.9	55.2	58.1	60.4	57.3
Control (unvaccinated plants)					
57.4	52.3	52.8	53.5	58.4	54.9

In seasonal dynamics, maximum values of leaf water deficit (Table 19) were observed in both varieties and unvaccinated plants. The deficit of leaf water saturation gradually increases in them, which is replenished at night, when water consumption for transpiration sharply decreases.

Table 19

Water deficit of 6-year-old pistachio as a function of variety.

Water deficit, %					
May	June	July	August	September	Seasonal average
"Mountain Pearl" variety					
11.2	18.5	24.2	25.8	26.7	21.3
The Orzu variety					
12.1	20.2	23.1	25.6	27.3	21.6
Sort "Albina"					
12.3	20.3	22.5	23.4	25.7	20.9
Control (unvaccinated plants)					
14.9	18.3	20.4	20.8	21.4	19.1

The revealed regularity in the dynamics of physiological processes in 3-5-year-old pistachio plants grown on gravelly lands at the stationary site of Kokand LPS (in transpiration intensity, leaf humidity, deficit of leaf water saturation) is clearly traced in older 6-7-year-old pistachio plants grown on gravelly lands on the territory of Kokand forestry (site "Shoberdi").

Figure 5. Pistachio seedlings in a container

Moreover, this pattern is also characteristic of pistachio varieties first released in these forest conditions (Tables 17-19). Despite certain differences between the varieties in these indices, they in comparison with unvaccinated plants (control) develop well, characterized not only by higher transpiration intensity indices in daily and seasonal dynamics (Table 17), but also by higher obliquity indices (Table 20).) Thus, if in unvaccinated plants the average area of one compound leaf is 87.8 cm^2, then in varieties "Gornaya pearl", "Orzu" and "Albina" it is on average 149.8 cm2, with the maximum value of 178.6 cm2 - in variety "Gornaya pearl" with the average weight of one leaf 1.54 g, with the maximum value in variety "Gornaya pearl" - 1.76 g (Table.20).

Table 20

Data on area and mass of complex leaf of pistachio depending on variety

Sort	Area of one complex sheet, cm2	Average weight of one compound leaf, g

Mountain pearl	178.6	1.76
Orzu	119.7	1.40
Albina	151.3	1.46
Average by variety	149.8	1.54
Control (non-grafted plant)	87.8	1.10

As the data of Table 21 show, the average water consumption for transpiration depending on the variety during the vegetation period is more than 52 mm, with leaf mass per one grafted tree 585.8 g.

Table 21

Water consumption for transpiration by young pistachio plants depending on variety during the vegetation period under their cultivation on gravels in the Fergana Valley.

Average seasonal transpiration rate, mg/g/hour	Leaf weight of 1 plant, g	Number of hours, days, months per growing season, hour	Number of plants per 1 ha, pcs.	Water consumption for transpiration per season from 1 ha, mm
"Mountain Pearl" variety				
2321,0	750	1980	210	72,3
Sort **"Albina"**				
1945,3	566,4	1980	210	45,6
The Orzu variety				
2162,9	441,0	1980	210	39,6
Average water consumption for transpiration by varieties.				
2143	585,8	1980	210	52,5
Control (non-grafted plant)				
2429,2	264	1980	210	26,6

Thus, pistachio, even under cultivation in very harsh forest conditions on pebble lands, practically in the absence of fertile soil horizon, retains the dynamics of physiological processes inherent to this species. It possesses at a young age good oblongness and, consequently, intensive assimilation activity and, therefore, can be recommended for cultivation of industrial plantations in the Fergana Valley.

Table 22

Moisture content of leaves of 5-year old pistachio depending on variety

Leaf moisture in, % but wet weight					
May	June	July	August	September	season average
Mountain pearl					
58.7	55.6	56.1	58.8	61.2	58.0
Orzu					
58.9	55.0	53.6	57.0	60.5	57.0
Albina					
56.2	53.6	53.9	57.0	59.8	56.1
Control (non-grafted plant)					
56.3	51.7	51.6	52.9	57.5	54.0

Relatively stable moisture content in leaves during the growing season is an important physiological feature of pistachio, which allows this plant not to lose leaf turgor even in the hottest and driest summer time.

Due to the dynamics (seasonal, daytime) of transpiration intensity decrease from spring to fall, the water saturation deficit gradually increases in pistachio, which is replenished at night, when water consumption for transpiration sharply decreases (Table 23).

Table 23

Water deficit of 6-year old real pistachio depending on assortment

Water deficit, %					
May	June	July	August	September	season average
Mountain pearl					
10.8	17.6	22.7	23.3	25.2	20.1
Orzu					

11.8	19.4	22.4	24.5	26.0	20.8
Albina					
12.1	19.5	21.8	22.2	24.5	20.0
Control (non-grafted plant)					
14.1	17.0	19.3	19.5	20.3	18.1

9.1 Study of daily and seasonal dynamics of water regime (transpiration rates, water deficit, leaf wetness) of varieties to establish the response of pistachio to new growing conditions.

As observations on water regime of young (3-6 years old) pistachio plants grown on pebble lands, on the territories of Kokand forestry (site "Shoberdi" foothills of Turkestan ridge) and Kokand LOS (foothills of Alai ridge), conducted in the period 20092011, have shown.The pistachio has shown high adaptation to the harsh forest conditions characteristic of the zone of gravel formation in the Fergana Valley. Practically in the absence of fertile soil horizon high summer air temperatures, low relative air humidity, intensive wind activity, pistachio varieties in new conditions retained inherent in this species high daily and seasonal transpiration intensity, almost constant water content in leaves and relatively insignificant water saturation deficit in leaves (Table 24).

Table 24.

Transpiration intensity of real pistachio in seasonal dynamics of variety dependence

Transpiration intensity, mg/g/hour					
May	June	July	August	September	Seasonal average
Mountain Pearl					
1988.2	2348.2	2917.3	2524.3	2248.2	2405.2
The Orzu variety					
1728.2	1953.4	2787.8	2629.3	2275.6	2274.8

Albina variety					
1988.4	2009.5	2568.3	2045.6	1568.5	2036.1
Control (non-grafted plants)					
2521.4	2828.2	3112.5	2254.8	1826.8	2508.7

Table 25.

Moisture content of leaves of 6-year old pistachio depending on variety

Leaf moisture in % per raw weight					
May	June	July	August	September	Seasonal average
Mountain Pearl					
59.2	56.7	57.2	59.3	62.4	58.96
The Orzu variety					
59.8	56.1	54.8	58.5	61.4	58.12
Albina variety					
57.9	54.9	55.2	58.1	60.4	57.3
Control (non-grafted plants)					
57.4	52.3	52.8	53.5	58.4	54.9

Table 26.

Water deficit of 6-year-old pistachio as a function of variety.

Water deficit, %					
May	June	July	August	September	Seasonal average
Mountain Pearl					
11.2	18.5	24.2	25.8	26.7	21.3
The Orzu variety					
12.1	20.2	23.1	25.6	27.3	21.6
Albina variety					
12.3	20.3	22.5	23.4	25.7	20.9

Control (non-grafted plants)					
14.9	18.3	20.4	20.8	21.4	19.1

As can be seen from the data presented in Table 24, transpiration intensity of both varieties and grafted seedlings increases from spring to summer, reaching its maximum in July-August. According to I.N. Beydeman (1960), this rhythm of transpiration intensity, which increases by summer and decreases by fall, is peculiar to plants constantly supplied with moisture. According to I.N. Beydeman's classification of three rhythms of transpiration intensity, in Fergana Valley conditions on gravelly lands, pistachio can be referred to the first type of seasonal rhythm, i.e. transpiration intensity increases until mid-summer, falling by fall.

Real pistachio is characterized by a relatively high and stable water content in leaves, which, as a rule, does not fall below the 50% mark. The same regularity was observed both in varieties and in unvaccinated plants, when leaf moisture varied on average during vegetation from 57.5% to 59.4% in varieties, and from 52.3 to 58.4% in control unvaccinated plants.(Table 25).

In seasonal dynamics, maximum values of leaf water deficit (Table 26) were observed in both varieties and unvaccinated plants. The deficit of leaf water saturation gradually increases in them, which is replenished at night, when water consumption for transpiration sharply decreases

Thus, the real pistachio, even under cultivation in very harsh forest conditions on zakolmatized pebble lands, retains the dynamics of physiological processes inherent to this breed. This regularity is manifested in the first 3 zoned varieties of pistachio. This allows us to predict the prospects and possibility of inclusion of other varieties of pistachio, recommended during the implementation of works under this project, into the composition of the released assortment.

CHAPTER 10. Biological resistance of pistachio (Pistacia vera L.) through the features of carbon metabolism

The study of carbohydrate metabolism is very important in determining the adaptation of plants to environmental conditions. Carbohydrates are products of assimilation activity of leaf mass, i.e. photosynthesis. The intensity of photosynthesis determines the reaction of plants to the conditions of their growth, including the conditions of moisture supply.

It has been established (Nasyrov, 1962; Akhmatov, 1976) that xeromorphic plants, which include Pistacia vera L., are characterized by prolonged and stable assimilation activity during the whole vegetation period. At the same time, Pistacia vera L. exhibited a high value of the average seasonal photosynthesis intensity (32-38 mg CO2/dm2 per hour), found at a temperature of 31-32°C and illumination of 60-64 thousand lux. This indicates the exceptional stability and adaptive capacity of the pistachio assimilation apparatus to the action of high temperature and illumination.

It is widely known that soluble sugars have a role in plant stability and adaptability.

For example, the role of sugars in frost resistance has been well and fully disclosed (Maximov, 1929, 1952; Tumanov and Trunova, 1957).

There is also data on plant resistance and adaptation (Suslova, 1941; Nekrasova 1949). These authors found that the content of soluble sugars in plant leaves increases in the heat of summer in different ecological types to a different extent.

According to Troshin (1956), sugars adsorbed on protein micelles have a stabilizing effect and protect them from the influence of various coagulating factors (high temperature, salt hypotony). Y.G. Molotkovsky, I.M. Zhestkova (1964), note that sugars stabilize respiration, making it relatively insensitive to overheating and respiratory poisons. According to the authors' conclusion, sugars block the active surfaces of mitochondrial membranes, protecting them from external influences.

From the above, the role of soluble sugars in plant resistance to overheating and dehydration, is quite multifaceted.

According to P.P. Chuvaev, N.A. Semenova, A.M. Shirshoeva (1962), in absolute accumulation of reserve substances (carbohydrates+ fats) in annual shoots of almond (Amygdalus communis L.) and pistachio (Pistacia vera L.) at the end of July, in the hottest time of summer (when the air temperature rises to 37-40°C), in the conditions of the Vakhsh valley (Tajikistan), there is no depression - accumulation of organic matter, but on the contrary goes continuously throughout the daylight hours.

Carbohydrates are known to serve as the main source of energy for metabolism and reserve substances, they are also necessary for increasing body mass and for the formation of fruits and seeds.

The main carbohydrate reserves of woody species are found in leaves, trunk and roots. The concentration of carbohydrates in individual plant parts varies (the highest concentration is in leaves and roots).

Soluble sugars (mono and disaccharides) serve as carbohydrate transport. Hemicellulose and starch serve as reserve carbohydrates. Starch is the final product of the assimilation process in higher plants.

It should be noted that there is a close relationship between the amount of starch in leaves and the adaptability of woody plants.

Adapted species (Armenica pinnatoramosa, Fraxinus lanceolata) are characterized by somewhat reduced synthetic capacity of leaves during wet periods and strongly weakened starch-accumulating activity during drought.

According to V.A. Kumakov (1952), starch supply of trees in autumn is as if the final expression of carbohydrate balance for the elapsed vegetation period, and characterizes the preparation of plants for winter.

Thus, carbohydrates are the main source of energy for plant metabolism, their reserve substances. The level of their content in different parts of plants (leaves, trunks, roots) determines the level (activity) of biochemical reactions in plants and, consequently, their response to growing conditions.

Since mono and disaccharides act as carbohydrate transporters, and starch and hemicellulose are the end products of assimilation.

Determination of carbohydrate metabolism in leaves of studied pistachio varieties in seasonal dynamics, from May to September inclusive, soluble and reserve carbohydrates were determined in the laboratory of UzNIILKh according to the abbreviated scheme of A.R. Kiesel (1934) with the use of micromethod of N. Bierry (Bierry H., 1951).

Non-grafted rootstocks of pistachio identical in age with the varieties served as a control. Determination of carbohydrate metabolism in leaves of the studied pistachio cultivars was carried out in seasonal dynamics from May to September inclusive during three years (2009 - 2010). Three varieties (Albina, Orzu, Gornaya pearl), zoned for rainfed zones, were included in the trial. 10-15 model plants and a control (non-grafted plant) were selected.

Analysis of the data presented in Table 27, as well as seasonal dynamics of carbohydrate accumulation shows that regardless of the individual characteristics of varieties and according to the indicators in the control plants, the content of soluble sugars (mono + disaccharides) and reserve nutrients (starch + hemicellulose) in pistachio leaves increases from early summer to autumn. Thus, if at the beginning of summer (May) in these three varieties the content of mono + disaccharides = from 4.77 to 5.31%, and starch + hemicellulose from 10.85 to 14.37%, then by autumn these indices increase almost 2 times and correspond for mono and disaccharides from 9.87% to 11.24%, and starch + hemicellulose - from 20.43% to 23.32%. Especially intensive carbohydrate metabolism is possessed by the named varieties - Orzu, Mountain Pearl and Albina, in which the sum of sugars by fall increased from 17.58% to 31.95%. These pistachio varieties are characterized by high content of soluble and reserve sugars, which indicates intensive carbohydrate metabolism and good adaptation (adaptation) of the tested variety in these soil and climatic conditions of the study area. The high content of reserve sugars (starch+ hemicellulose) in the named varieties also contributes to a more intensive adaptation (adaptation) to local environmental conditions, characterizing its drought resistance and since all this occurs at very high summer air temperatures +37° and +40°C there is an intensive process of formation of reserve sugars (starch+ hemicellulose) from July to September.

According to V.A. Kumakov (1952), starch availability of trees in autumn is the final expression of carbohydrate balance for the past vegetation period and characterizes the preparation of woody plants for winter, i.e. their winter hardiness. The growth of woody plants in height and phloem activity are closely related to fluctuations in the amount of starch. If the maximum of starch accumulation occurs in the fall, the time when phloem begins to differentiate.

According to modern views, the protective value of sugars is to stabilize the structure and function of important macromolecules during the action of adverse environmental factors.

Carbohydrate regime plays a leading role in adaptation of plants to unfavorable environmental conditions.

The obtained data on accumulation in three varieties of true pistachio confirms its important diagnostic value in the selection of the best varieties to drought, i.e. its drought resistance increases. Intensive accumulation of reserve sugars of pistachio by the end of vegetation, contributes to more intensive preparation for winter, i.e. increases its winter hardiness.

Thus, the high level of accumulation of soluble and reserve carbohydrates in the leaves of pistachio leaves of the tested varieties (Albina, Orzu and Gornaya pearl) indicates the stability and development of adaptive properties of the tested variety to these growing conditions.

Carbohydrate metabolism plays a leading role in the adaptive ability of plants to unfavorable environmental conditions, which in these conditions are high air temperatures, high light and strong wind activity in the hot period (July). The obtained data on the accumulation of starch and hemicellulose in pistachio confirm its important diagnostic value in the selection of the best varieties to drought, i.e. its drought tolerance increases. The accumulation of reserve sugars in pistachio and their sharp increase by the end of the growing season contributes to more intensive preparation of the plant for winter, i.e. increases its winter hardiness (up to -35-40°C).

Thus, the level of accumulation of soluble and reserve carbohydrates in the leaves of pistachio of the three varieties indicates stability and good adaptive capacity to arid harsh growing conditions on gravelly soils in the Fergana Valley.

Table 27

Seasonal dynamics of accumulation of soluble and reserve sugars in % per absolute dry mass in leaves of
Pistacia vera L. Average for 2008-2010.

Experiment variant (variety)	May		June		July		August		September	
	Mono + Disaha ra, M m±	Starch+ Hemicellulose vine, M m±	Mono + Disaha ra, M m±	Starch+ Hemicellulose vine, M m±	Mono + Disaha ra, M m±	Starch+ Hemicellulose vine, M m±	Mono + Disaha ra, M m±	Starch+ Hemicellulose vine, M m±	Mono + Disaha ra, M m±	Starch+ Hemicellulose vine, M m±
Mountain pearl (C20	5.25± 0.04	14.37± 0.08	6.45± 0.07	16.06± 0.05	7.87± 0.07	18.02± 0.04	8.87± 0.05	19.27± 0.08	11.24 ± 0.05	23.22± 0.08
Orzu (C-172)	5.31± 0.03	12.18± 0.04	6.70± 0.04	14.44± 0.06	7.88± 0.06	16.55± 0.02	8.51± 0.04	17.22± 0.07	10.57 ± 0.08	21.65± 0.06
Albina (C-11-"A")	4.77± 0.03	10.85± 0.07	5.89± 0.06	12.95± 0.03	7.18± 0.05	14.98± 0.06	7.35± 0.09	16.23± 0.02	9.87± 0.03	20.43± 0.04
Average by variety	5,11	12,46	6,34	14,48	7,64	16,52	8,24	17,57	10,56	21,8
Control 1 (ungrafted tree)	3.75± 0.04	11.21± 0.07	4.85± 0.05	12.35± 0.07	6.00± 0.03	14.8± 0.04	6.03± 0.07	15.18± 0.09	7.25± 0.07	16.28± 0.08

10.1 Description of studied varieties of true pistachio (collection of G.M.

Chernova).

"Mountain Pearl" variety

The tree is medium-sized, with a loose rounded crown. It enters economic fruiting from 12-15 years of age. Fruits (nuts) are single-seeded bones gathered in compact rounded brushes (12-15 pieces in a brush). The pericarp of the bones (nuts) is white by the time of fruit ripening. Nuts are large, 19x13x12 mm, one-dimensional, rounded, slightly ribbed. Shell cracking at the seams is bilateral, at 4/5 of the seam length. Shell thin, slightly rough, grayish. The shell of the kernel is light pink, the pulp of the kernel is dry, dense, pistachio-colored. The taste of the kernel is sweetish.

MOUNTAIN PEARL

Figure 6. Fruits of the pistachio variety "Mountain Pearl"

The variety is late maturing (in conditions of Southern Tajikistan and rainfed foothills of Uzbekistan - third decade of August - first decade of September). Yield up to 8 kg per 1 ha. Pollinator 20-4; 11 "A"-3. Medium resistant to fungal diseases and fruit damage by pests. Resistant to the effects of atmospheric droughts (dry winds). Medium-demanding to soil fertility.

The yield of cracked nuts from the total crop weight is 90%, well transportable. The kernel does not lose its flavor during 2-3 years of storage. It is characterized by a high content of sugars, up to 5%. Fat content-57%, protein-13%.

The recommended plantation density is 180-200 trees per 1 ha.

Sort "Albina"

Medium-sized tree with a wide-spreading openwork crown. Fruits (nuts) are single-seeded bones gathered in loose elongated (up to 16-20cm) brushes. On average 25-30 nuts in a brush.

ALBINA

Figure 7: Fruit of pistachio variety "Albina"

The pericarp (nuts) is whitish by the time of fruit ripening, light pink at the tip. Nuts are medium-sized, 17x10x9mm, medium unidimensional, elongated-elipsoid shape, slightly ribbed. Shell cracking along the seams is mostly unilateral, ½ the length of the seam. Shell thin, slightly rough, light-colored. The shell of the kernel is light pink, the pulp of the kernel is dry, dense, light green in color. The flavor of the kernel is sweet.

The variety is medium maturing (in conditions of Southern Tajikistan - the first - beginning of the second decade of August). Yield up to 8 tons per 1ha. Pollinator 11 "A" -3. Resistant to fungal diseases and fruit damage by pests. Exceptionally resistant to the effects of atmospheric droughts (dry winds). Medium-demanding to soil fertility.

The yield of cracked nuts from the total crop weight is 80-85%. Well transportable. The kernel does not lose its flavor during 3 years of storage. The kernel is characterized by an increase in the content of sugars, up to 5%. Fat content - 59%, protein - 13%.

The Orzu variety

The tree is strong-growing, with a wide-spreading rounded crown, It enters economic fruiting from 10-12 years of age. Fruits (nuts) are single-seeded bones collected in compact, medium-length (8-10 cm) brushes. On average, 15-18 pieces in a brush. The pericarp of the bones (nuts) is white by the time of fruit ripening.

ORZU

Figure 8. Fruits of pistachio variety "Orzu"

Nuts are large, 19x14x13 mm, medium unidimensional, ellipsoidal in shape, slightly silvered. Shell cracking at the seams is bilateral, at 3/4 of the seam length. The shell is thin, slightly rough, white to grayish in color. The shell of the kernel is dark pink, the pulp of the kernel is dry, dense, pistachio (light green) in color. The flavor of the kernel is slightly sweet.

Early maturing variety (in conditions of Southern Tajikistan and rainfed foothills of Uzbekistan the beginning of the first decade of August). Yield up to 10 c/ha. Attributed to the category of "high-yielding". Pollinator 172-3, 11 "A" -3. Exceptionally resistant to fruit damage by pests. Resistant to the effects of atmospheric droughts (dry winds). Responds well to feeding with organic-mineral fertilizers.

The yield of cracked nuts from the total crop weight is 75-80%. Well transportable. The kernel does not lose its flavor during two years of storage. Kernel content of sugars-4%, fat-59%, protein-15%.

Sort - dessert, for the food industry.

CONCLUSION

The substantiation of biological and ecological adaptive capacity of the pistachio allows us to hope for intensification of involvement in agricultural turnover of many empty rainfed foothills and low mountains by cultivation of plantation (garden) pistachio crops.

The true pistachio is a valuable natural phenomenon in the arid zone of southern Central Asia. Preservation and replenishment of this species in its ancestral homeland - in mountainous and foothill areas - is of great conservation importance for the whole Central Asian region.

The main environmental factor limiting the condition of pistachio is temperature regime and soil moisture.

For normal growth and development of pistachio throughout the whole region of its growing in Central Asia, the sums of stable average daily air temperatures above +5^{0}C not less than 3400$^{(0)}$; above +10$^{(0)}$C not less than 3200; above +20^{0}C not less than 2000-2200$^{(0)}$C are necessary. The average duration of the vegetation period should be 200 or more days, and frost-free period - not less than 160 days.

The wide geographical and altitudinal range of pistachio growth, first of all, testifies to the high adaptation of this breed, which allows it to grow in various natural and climatic conditions of the Central Asian region and, consequently, to the unlimited territorial possibilities of its introduction into culture.

A biological and morphological feature of pistachio is the development of a powerful root system, which already in the first year grows and reaches a depth of 100 (150) cm, and then horizontal roots with a huge network of small sucking roots are already developing. Adaptability of pistachio to develop a powerful root system in different soil and climatic conditions allows to use moisture and nutrients from a large area, which is one of the main adaptive properties.

The study of the peculiarities of water regime of pistachio is an indicator of resistance and adaptation to drought. Biological adaptation of drought-tolerant plants, including the true pistachio, is primarily conditioned by resistance to dehydration, which is reflected in their water regime.

The development of a powerful root system of pistachio provides throughout the growing season a high index of transpiration intensity, which in turn increases assimilation activity in contrast to other tree species and is characterized by a long duration during the growing season, due to wide open stomata during the summer period.

The true pistachio is characterized by increased transpiration intensity, which correlates primarily with air temperature. Prolonged overheating of leaves causes deep disturbances in metabolism, with transpiration acting as the main physiological factor stabilizing leaf temperature.

Therefore, in order to establish the high adaptive capacity of pistachio, which is confirmed by the wide geographical and altitudinal range of natural distribution of this species within Central Asia, in the period 2006-2012, studies were conducted on the prospects and possibility of pistachio cultivation in very harsh natural-climatic conditions characteristic of pebble lands in the Fergana forest region of Uzbekistan.

The study of soil environment and content of basic elements of soil nutrition (nitrogen, phosphorus) showed that soils by granulometric composition are dusty-clayey with transition to pebbles. They are low-structured and difficult to be cultivated. In terms of humus, gross nitrogen and phosphorus content soils are not characteristic and can be referred to pebble soils.

Therefore, carrying out a set of agro-management practices, including vegetation furrow irrigation (5-6 per vegetation), as well as the application of film-protective

preparations ("meliorant") improve conditions of moisture availability of pistachio, providing more favorable conditions for their growth and condition on zakolmatized gravels, typical for the Fergana forest region.

Thus, the real pistachio, even when grown in very harsh forest conditions on zakolmatized pebble lands, retains the dynamics of physiological processes inherent to this breed. This regularity is manifested in the first 3 zoned varieties of pistachio. This allows us to predict the prospects and possibility of inclusion of other pistachio varieties recommended in the composition of the released assortment of this project.

The study of carbohydrate metabolism is very important in determining the adaptation of plants to environmental conditions. The level of accumulation of soluble and reserve carbohydrates in the leaves of pistachio of the present three varieties indicates stability and good adaptation to arid habitat conditions on gravelly lands in the Fergana Valley.

USED LITERATURE

1. Abu-Ali Ibn Sina. Canon of Medical Science Volume 2, ed.2.-Tashkent, "Fan", 1982, pp.29-48.

2. Adamovich Z.I., Norin B.N. The resin-bearing system of the true pistachio Pistacia vera L.-"Botan.zhurn.", 1956,№ 9.

3. Alexeev V.P. Pistachia vera L. Family Anacardiaceae // Subtropical Crops. 1963. № 3.C.85-92

4. Alekseev A.I., Gusev N.A. Physiological analysis of plant water deficiency. In Sb.: Physiology of adaptation and stability of plants at introduction. Novosibirsk, "Nauka", SOANSSR, 1969.

5. Altergot V.F., Mordkovich S.S. Role of increased temperature in the complex effect of drought on plants. In Collect: Physiology of plant adaptation to soil conditions. "Nauka", SOANSSR, 1973.

6. Akhmatov K.L. Peculiarities of transpiration of forest-forming tree species of Chatkal ridge. Proceedings of the Kyrgyz Forest Experimental Station, Vol. 3, 1962.

7. Akhmatov K.A. Transpiration and temperature regime of leaves of tree and shrub species in the foothills of the Kyrgyz Alo-Too. Theses.dokl.Sh Uralskogo sobosch. po physiol. i. ekol. Woody plants. -UFA, 1970.

8. Bulychev A.S. Bioecological features of pistachio in the foothills of the Kyrgyz ridge. -Frunze, publishing house of the Academy of Sciences of the Kyrgyz SSR, 1969, 81 pp.

9. Bierry N., et al. (Bierry N.,) Compt.rend biolad,1951.

10. Beydeman I.N. - To the methodology of studying the water regime of plants. "Bot. zhurn.", T. XI. M. Nauka, 1956, No. 2, p. 212-220.

11. Dvoretskaya E.I. - To the question of drought resistance of summer oak and some other species. "Lesnoe khozyoz-vo" No. 2, 1949.

12. Vasilevskaya V.K. Leaf formation of drought-resistant plants - Ashgabat,

publishing house of the Academy of Sciences of Turkmen SSR, 1954.

13. Water regime and drought resistance of plants. Bibliogr. Index.-Kishinev, 1964. Izd. "Kartia Maldoveniaske".

14. Genkel P.A. Plant resistance to drought and ways to increase it. Proc. Institute of Plant Physiol. USSR Academy of Sciences, т.V, vol. I,1946, 236 pp.

15. Zheltikova T.A. Afforestation on pebble lands. Moscow: Izd. "Lesnaya Promyshlennaya" . 1971, c.165.

16. Zholkevich V.N. Respiration energetics of higher plants under conditions of water deficit. M., "Nauka". 1968.

17. Zapryagaeva V.I. Wild fruits of Tajikistan. M., L. 1964. p.694.

18. Ivanov L.A. About the method of determination of transpiration on cut shoots.-Bot.zhurn., 1956a, vol.44, No.2, p.219-220.

19. Ivanov L.A., Silina A.A., Celniker Y.L. - On the method of rapid weighing for determining transpiration in natural conditions. "Bot. zhurn.", 1950. Vol. 95, No. 2, pp. 171-185.

20. Keller B.A. Osmatic pressure and water content in assimilating organs of plants as means for elucidation and evaluation of their ecological-physiological and ecological-morphological peculiarities. -Plant and environment.-M.: Izd. of the Academy of Sciences of the USSR, 1952, vol.3, p.171-212.

21. Kushnirenko M.D. Water regime and drought resistance of fruit plants.- Kishinev, Shtiyintsa, 1962, 48 p.

22. Kenesarina N.A. Transpiration of tree and shrub species in Karaganda conditions. In book: Scientific conference on rationalization of forestry and agroforestry in Kazakhstan.-Alma-Ata, 1959.

23. Kokina S.I. Water regime and internal factors of plant stability in the sandy desert of Kara-Kum. In Collected Works: Problems of plant breeding development of deserts, v.4, 1935.

24. Kondo I.N. Water regime of grapes under drought conditions. Proc. In-ta physiol. plants ANSSR, т.IV, v.1, 1946.

25. Kiesel A.R. A practical guide to plant biochemistry. Biomediz, 1934.

26. Kumakov V.N. Dynamics of reserve carbohydrates of oak, ash and elm in connection with their growth and stability in the south-east. Avt.kand.diss.- Saratov, 1952.

27. Labutin V.K. Sketches of adaptation in biology and technology.-L. "Energia", 1970.

28. Lebedintseva E.V. Physiological and anatomical features of plants grown in dry and humid atmosphere. Izd. of the Gl. bot. garden, т.XXV, vol. 4, 1926.

29. Lisitsina G.N. Natural environment of the South of Central Asia in the Holocene. The Stone Age of Central Asia and Kazakhstan. Tashkent, "Fan", 1972.

30. Maksimov N.A. Selected works on drought resistance and winter-hardiness of plants. T.I. Water regime and drought resistance of plants. Moscow: Izd-voor ANSSR, 1952, 576 p.

31. Massalsky V.N. Turkestan region, St. Petersburg, 1913. 861 c.

32. Matveev M.I. Water rheem of some woody plants of mountainous Tajikistan. Stalinabad. Publishing house of the Academy of Sciences of the Tajik SSR, 1953, 84 p.

33. Mikhailova L.M., Tuliaganov T.E. Adaptation of pistachio real to different natural-climatic environmental conditions in connection with cultivation on pebble lands in the Fergana Valley. AGRO ILM 2011.

34. Mikhaylova L.M., Chernova G.M., Rakhmonov A.M. Assessment of Resilience of pistachio (Pistacia vera L.) through the features of carbohydrate metabolism and prospects for the use of valuable variety on leased plots. Republican Scientific and Practical Conference "Conservation and sustainable use of biodiversity of agricultural crops and their wild relatives" December 10, 2009, Tashkent, Uzbekistan. C.98-101.

35. Nasyrov Y. S. Photosynthesis of mesophilic and xerophilic tree species of the Kondara Gorge. Proc. Department of physiology and biophysics of plants. Academy of Sciences of Tajikistan. SSR, vol. 1, 1962a, 18-35.

36. Nikoliai L.V. New effective technology of growing pistachio plantations in Uzbekistan. AGRO ILM 2011.

37. Petinov N.S. State and prospects of water regime of plants in the USSR. In Collected Works: Water regime of agricultural plants. Moscow, "Nauka", 1969.

38. Petinov N.S., Molotkovsky Y.G. Defense reactions of heat-resistant plants under the action of high temperatures. "Physiology of Plants, Vol. 4, Issue 3, 1957.

39. Popov K.P. Pistachio in Central Asia. Ashgabat. 1979. c.159.

40. Popov K.P. On the distribution of fine roots of pistachio along soil profiles.- "Lesovedenie" 1971a № I.

41. Popov K.P. On vegetative renewal of pistachio real.-"Lesovedenie" 1974a, [1] I.

42. Satarova N.A. Some regulatory mechanisms of plant adaptation to drought and high temperatures. In Collected Works: Physiology of drought resistance of plants. Moscow, "Nauka" 1971.

43. Sisakyan N.M., Kobyakova A.M. Directionality of enzymatic action as a sign of drought resistance of cultivated plants. Report. VI Adsorption of invertase by plant tissues during wilting. "Biochemistry",1947, vol. 12. № 5.

44. Suslova M.I. Coefficient of wilting of pistachio and almonds of the Western Kopetdagh. Dokl. VASKhNIL, v.21, 1940, p.6-9.

45. Chernova G.M., Olekhnovich G.S. About water regime of pistachio in forest-garden crops in the south of Tajikistan. Forest Science. -M.: ANSSR, 1975, No. 2, p. 64-69.

46. Shcheglov V.F., Nimadjanova K.N., Babekova E.Y., Vakhobova H.V. Dynamics of carbohydrate accumulation in the kernels of pistachio nuts. // Proceedings of the Academy of Sciences of the Taj.SSR. 1975. T.59. №2. C.70-76.

47. Avanzato D, Monastra F. - Coltura del pistachio: Situazione attuale e ricerche in cors. - Fruitticoltura, 1981, 43. 10-11, 15-18.

48. Avanzato D, Monastra F. - Coltura del pistachio: Situazione attuale e ricerche in cors. - Informatore di ortotiolofruitti-coltura 1982, 23, 6:1519.

49. Ayfer M. La culturedupistachier enTurguie. "Fruits, vol. 22,№ 8, 1967, Paris.

Printed by Books on Demand GmbH, Norderstedt / Germany